KS2 Success

Revision Guide

Paul Broadbent

Maths
SATs

Contents

Numbers

Place value 4
Comparing and ordering 6
Rounding and approximation 8
Number patterns 10
Formulae and equations 12
Special numbers 14
Factors and multiples 16
Fractions 18
Comparing fractions 20
Decimals and fractions 22
Comparing decimals 24
Ratio and proportion 26
Percentages 28
Test your skills 30
Test your knowledge 32

Calculations

Mental maths 34
Multiplication facts 36
Addition and subtraction 38
Multiplication methods 40
Division methods 42
Fractions of quantities 44
Percentages and money 46
Problems 48
Test your skills 50
Test your knowledge 52

Shapes

2-D shapes	54
3-D solids	56
Geometry	58
Using coordinates	60
Angles	62
Test your skills	64
Test your knowledge	66

Measures

Measures	68
Time	70
Area and perimeter	72
Test your skills	74
Test your knowledge	76

Data

Probability	78
Charts and graphs	80
Line graphs	82
Averages	84
Test your skills	86
Test your knowledge	88

National Test practice

National Test practice	90

Answers and glossary

Answers	93
Glossary	95

Place value

Large numbers

The position of a **digit** in a number is really important. In a race, 912th is a long way behind 129th, even though the digits are the same.

The **millions** and **thousands** in large numbers are separated by gaps or commas. **When you read out large numbers, say the words million or thousand at these gaps.**

				5	3	fifty-three
			2	9	0	two hundred and ninety
		4	1	0	6	four thousand, one hundred and six
	3	8	6	1	9	thirty-eight thousand, six hundred and nineteen
1	5	0	6	3	8	one hundred and fifty thousand, six hundred and thirty-eight
2 7	1	4	1	3	0	two million, seven hundred and fourteen thousand, one hundred and thirty

Example:

Writing 31,425 in words.

Put a space or comma after the thousands digit and read it like separate numbers.

31 425

Thirty-one thousand, four hundred and twenty-five

Digits

Ten digits are used to make all our numbers: 0 1 2 3 4 5 6 7 8 9

Example: 3 7 0 9 4

- Make the largest number possible from these five digits.
- Make a number as near as possible to 50 000 from these five digits.

For questions like these, **just start at the ten thousands and work your way down the numbers**, rearranging the digits as you go.

Largest number → 97 430

Nearest to 50 000 → 49 730

Zero is a very important digit, especially when you make a number ten or a hundred times bigger. Zeros are then used to fill the spaces.

× and ÷ by 10 and 100

Making a number ten or one hundred times bigger or smaller is easy, as long as you remember these rules:

To multiply by 10:

Move the digits one place to the left and fill the space with a zero

3 8 6
3 8 6 0 ×10

To multiply by 100:

Move the digits two places to the left and fill the two spaces with zeros

4 1 8
4 1 8 0 0 ×100

To divide by 10:

Move the digits one place to the right

4 7 0
4 7 ÷10

To divide by 100:

Move the digits two places to the right

3 6 0 0
3 6 ÷100

To see what happens with decimals, turn to page 23.

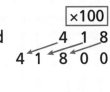
PLACE VALUE NUMBERS

Life would be very different if we didn't use zeros carefully.

Yes - Mum might give us £500 instead of £5 pocket money!

Have a go...

- Key a large number into your calculator. Multiply or divide that number by 10 or 100 and look at what happens.
- Explain to someone what happens to the digits of a number if you multiply or divide it by 10.

Key words

digit   thousand
million

Quick Test

1. Write the number 23 608 in words.
2. What is the smallest number you can make from the digits 3 7 1 8 4?
3. 1896 × 100
4. 34 500 ÷ 100
5. Write the number ten thousand and eighty-four in digits.

Comparing and ordering

Comparing the size of numbers

When you need to compare two numbers, it is helpful to **write them under each other**, lining up the units.

Example:

This centipede is 6214 mm long

This millipede is 6197 mm long

Which is the longer?

Line up the units.

Th thousands	H hundreds	T tens	U units
6	2	1	4
6	1	9	7

As you can see, 6214 is bigger – it has two hundreds.

< and >

These are really useful **symbols**... as long as you do not get them confused!

< means is less than

For example, 385 < 450

385 is less than 450.

> means is more than

For example, 450 > 385

450 is more than 385.

Can you think of a good way to remember these?

I think of them like a crocodile's open mouth. The open mouth always goes for the biggest number.

Ordering numbers

If you have a list of numbers to put in order, the first thing to do is to **put the numbers into groups with the same number of digits**.

Arrange each group in order of size depending on the **place value of the digits**.

Remember always to read numbers from the left to the right.

Example:

These are the results of the annual 'Bean-eating with a tooth-pick' competition.

Write them in order, starting with the smallest.

Name	No. of beans
Billy	704
Sandra	2145
Martin	19 045
Joe	689
Alice	3022
Fred	10 945
Mark	12 144
Sue	3689
Roger	2089
Ann	698

Write them in order, in groups, with the same number of digits:

689, 698, 704; 2089, 2145, 3022, 3689;

10 945, 12 144, 19 045

 Top Tip *If two numbers are very close, find the higher one by comparing the digits in each column, starting at the left.*

 Have a go...

Ask someone:

- *to give you a list of large numbers. Put them in order of size.*
- *to give you three numbers. Use the < and > signs with the numbers.*

 Key words

symbol

Quick Test

1. Which is the larger number: 61 049 or 60 194?

2. Put these numbers in ascending order:
   384   348   438

3. Place these numbers in order, starting with the smallest:

   4205   2114   485   2044
   379   2611   10 901

4. True or false?

   a 439 > 493   c 888 > 901

   b 604 < 640

Rounding and approximation

Estimating

We often **estimate** measurements, but estimating is also useful for checking sums. **Estimating is a bit like guessing** – but we use information to get a more accurate **approximate** answer, rather than a wild guess!

Rounding

Rounding makes numbers easier to work with – using tens, hundreds or thousands.

We round numbers for many reasons:
- if exact answers are not needed
- to help remember numbers
- to work out approximate answers before calculating.

≈ means 'is approximately equal to'

Rounding to the nearest 10:

Look at the units digit.

If it is 5 or more, round up the tens digit. If it is less than 5, the tens digit stays the same.

315 rounds up to 320 and 234 rounds down to 230.

Rounding to the nearest 100:

Look at the tens digit.

If it is 5 or more, round up the hundreds digit. If it is less than 5, the hundreds digit stays the same.

1453 rounds up to 1500 and 2648 rounds down to 2600.

Rounding to the nearest 1000:

Look at the hundreds digit.

If it is 5 or more, round up the thousands digit. If it is less than 5, the thousands digit stays the same.

13 854 rounds up to 14 000 and 49 389 rounds down to 49 000.

It's useful to be able to estimate things.

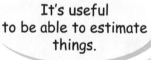

Like how many sweets you can get with your pocket money.

Approximate answers

When you carry out a calculation, **work out an approximate answer before you start**. Then you will know if your actual answer makes sense. Use rounding skills to work out approximate answers for these:

> There are 42 burgers in a box. How many are there in 29 boxes?

To get an approximate answer, simply use the nearest round numbers: 40 × 30 = 1200.

> There are 504 chicken nuggets in a bag. Each 'Cluckmeal' has 9 nuggets. How many 'Cluckmeals' will there be?

An estimate would be: 500 ÷ 10 = 50.

Sensible amounts

Newspaper headlines often round numbers to make them sound better. If a crowd of 50 329 go to watch a pop concert the headline might read:

> **50 000 screaming fans greet The Space Boys**

If £3 110 924 was won on the lottery, the headline could be:

> **Over £3 000 000 to spend, spend, spend!**

Fisherman lands a 21 840 g carp in garden pond!

What would be a good headline for this?

Have a go...

Look for numbers written in newspapers. Round them:
- *to the nearest 10*
- *to the nearest 100*
- *to the nearest 1000*

Key words

estimate   rounding
approximate

Quick Test

1. Round 3455 to the nearest 10.
2. Round 48 473 to the nearest 100.
3. Round 686 474 to the nearest 1000.
4. What is an approximate answer for 996 × 31?
5. Which of these numbers rounds to 45 500 to the nearest 100?
   - a 46 015
   - b 44 683
   - c 45 486
   - d 45 562

Number patterns

Number sequences

A **sequence** is a list of numbers which usually has a pattern. You can often find the pattern or rule in a sequence by looking at the **difference between the numbers**.

This is a typical test question:

Write the next number in this sequence.   19   15   11   7   _3_

It is a bit like cracking a code – look at the difference between the numbers. Each number is 4 less than the previous one, so the missing number is 3.

Try this one:

Write the missing number in this sequence.   3   6   _12_   24   48

Look at the way the numbers grow. They keep doubling, so the missing number is 12.

Fibonacci sequence

This pattern is named after a brilliant medieval mathematician, Leonardo of Pisa. His nickname was Fibonacci and he discovered this sequence:

1   1   2   3   5   8   13   21

Can you see how the pattern grows?

What do you think the next number will be?

Each number is the total of the previous two, so the next number is 34.

Maybe it's the number of pizzas Leonardo ate each week?

It's Leonardo of Pisa – not pizza, silly!

Negative numbers

Positive numbers are **above zero** and negative numbers are **below zero**.

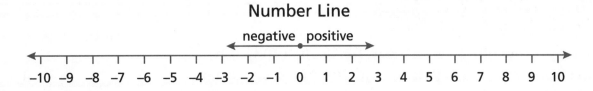

Some number sequences may include negative numbers:

−5 −3 −1 1 3 5

−7 −4 −1 2 5 8

Can you spot the patterns?

Look at the difference between each number.

Remember to include zero when you are looking at negative numbers in a sequence.

Top Tip: Draw 'jumps' between each number and write the differences. This might help you spot the pattern.

Reading thermometers

Test questions often use negative numbers on thermometers. **Read it like a number line** – just make sure you know what each number is on the scale.

By how many degrees has the temperature risen? To work this out, you need to calculate the difference between the two temperatures. The first one reads −9°C, the second reads 18°C, so the temperature has risen by 27°C.

Have a go...

Explain to someone:
- *the Fibonacci sequence*
- *how to read a thermometer.*

Key words

sequence negative number
positive number

Quick Test

1. What is the next number in these sequences?
   a 2 5 8 11
   b 12 7 2 −3
   c 231 222 213 204
   d 5 10 20 40

2. What is the difference between −8 and 3?

3. The temperature is 4°C. It drops by 7°C overnight. What is the night-time temperature?

Formulae and equations

Function machines

Function machines have three parts. You may have questions which ask you to find the numbers coming out of the machine, numbers going into the machine, or even the missing function.

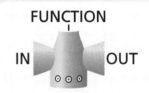

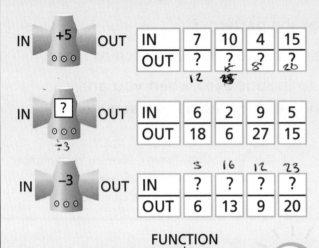

IN	7	10	4	15
OUT	? 12	? 15	? 8	? 20

IN	6	2	9	5
OUT	18	6	27	15

IN	? 3	? 16	? 12	? 23
OUT	6	13	9	20

If you are asked which number goes into the machine, you will need to know the **inverse** or opposite of the function.

- the opposite of adding is subtracting
- the opposite of multiplying is dividing

Top Tip *Some functions may have more than one operation. Just put a number in and carry out the operations in order: For example: $(6 \times 2) - 3 = 9$*

Formulae and equations

Example:
Tickets to the 'House of Horror' cost £2 each. What is the cost of n tickets? Cost = 2n

A **formula** (the plural is 'formulae') uses **letters or words to give a rule**.

Equations have **symbols or letters instead of numbers**.

$\boxed{4} + 2 = 6 \quad ? - 5 = 3 \quad 2s = 14$

This means 2 times s: the × sign for multiplication is not used in equations because it might look like a letter. That could confuse you!

You need to work out what the symbol or letter stands for.

Use the numbers to help you, and say it as a sentence: "What added to 2 makes 6?", "What minus 5 makes 3?" and "What times 2 makes 14?"

If you are finding it difficult to work out the value of a letter, **write it out again using a box instead of a letter**. You can then try different numbers in the box to see if the equation works.

$3t - 1 = 14 \qquad 3 \times \boxed{5} - 1 = 14$

$\qquad\qquad\qquad\qquad 3 \times \boxed{5} - 1 = 14$

$\qquad\qquad\qquad\qquad t = 5$

Tricky equations

If it is a complicated equation, you may need to **work it out step by step**:

$$4y + 3 = 15$$

1. You want y on one side of the equation, and the numbers on the other. Take away 3 from both sides:

   $$4y = 15 - 3 \quad 4y = 12$$

2. Say the equation as a sentence:

   4 times something makes 12.
   4 threes are 12.

3. Write the value of y:

   $$y = 3$$

4. Try it in the original equation to see if it works:

   $$4 \times 3 + 3 = 15$$

Try this one – remember the 4 steps:

$$2x - 5 = 7$$

1. Letters on one side – numbers on the other (for this one add 5 to both sides).
2. Say the equation.
3. Write the value of x.
4. Test it with the equation.

 Top Tip *Remember, equations need to stay balanced. If you add or take away a number from one side, do the same to the other side and the equation stays the same. It is a good way of working out the value of x or y.*

 Have a go...

- Draw some weird and wonderful function machines.
- Make up a rule or function for each of them.
- Write up a table of numbers going in and out of each machine.

 Key words

inverse   equation
formula

Quick Test

1. What are the missing numbers in this table?

   IN ×3 +1 OUT

IN	3	5	5	32
OUT	4	4	16	31

2. There are 20 gold rings on a tray. If n rings are sold, what equation shows how many are left?

3. $4t + 6 = 14$ What is the value of t?

4. Choose the correct function or rule for these numbers:

IN	1	2	3	4
OUT	3	5	7	9

   a ×2   c ×2 + 1
   b ×3 − 1   d +2

Special numbers

Odds and evens

All **even numbers** can be **divided exactly by 2**. You can recognise even numbers because they always end in **0, 2, 4, 6 or 8**.

132, 386, 14 030, 233 594… are all examples of even numbers.

An **odd number cannot be divided exactly by 2**. Odd numbers always end in **1, 3, 5, 7 or 9**.

427, 9355, 20 891, 344 863…

are all examples of odd numbers.

Top Tip: When you add and subtract numbers there are interesting rules for odds and evens. Check these: even + odd = odd; odd + odd = even; even + even = even; even − odd = odd; odd − odd = even; odd − even = odd.

Square numbers

Numbers multiplied by themselves make **square numbers**. They are called square numbers because they can be shown as squares (maths is so simple!).

A short way of writing 3 × 3 is 3^2, 4 × 4 is 4^2, 5 × 5 is 5^2 and so on…

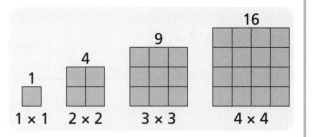

1 × 1, 2 × 2, 3 × 3, 4 × 4

Square roots

Square roots are the opposite of square numbers. To find the square root of, say, 25, find which number, when multiplied by itself, makes 25. So the square root of 25 is 5.

I wonder if there are rules for multiplying and dividing odds and evens?

I've come over all odd thinking about that!

Triangular numbers

Triangular numbers are numbers that are made by triangular patterns. Look at the examples below.

$1 + 2 = \boxed{3}$

$1 + 2 + 3 = \boxed{6}$

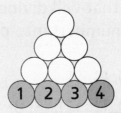
$1 + 2 + 3 + 4 = \boxed{10}$

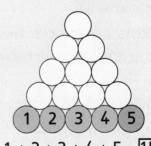
$1 + 2 + 3 + 4 + 5 = \boxed{15}$

Special patterns

Questions are often asked about the **patterns** that these special numbers make. You might be asked to write the **missing numbers in a sequence**.

Odd numbers

_1_, 3, 5, _7_, 9, _11_

Even numbers

2, _4_, 6, _8_, 10, _12_

Square numbers

1, 4, ____, 16, ____, 36

Triangular numbers

1, 3, 6, ____, 15, ____

Have a go...

Write down:
- the first 10 square numbers
- the first 10 triangular numbers.

Key words

even number   square root
odd number   triangular number
square number

Quick Test

$10 \times 10 = 100 = 10^2$

1. What is 10^2?
2. What is the fourth triangular number?
3. What are the missing numbers in this sequence?

    23 _21_ _19_ 17 15 13

4. What is the square root of 36?
5. What is the next square number after 100?

Factors and multiples

Factors

Factors are special numbers that will **divide exactly into other numbers**. It is useful to put factors of numbers into pairs:

Factors of 15 →	(1, 15), (3, 5)	→ 4 factors
Factors of 24 →	(1, 24), (2, 12), (3, 8), (4, 6) →	8 factors
Factors of 32 →	(1, 32), (2, 16), (4, 8)	→ 6 factors

Did you know that a number **always has an even number of factors** – unless it is a square number. Try it – how many factors do 9 or 16 have?

If a number only has 2 factors, itself and 1, then it is a **prime number**. For example, 17 is a prime number because it can only be divided exactly by 1 and 17. The first six prime numbers are 2, 3, 5, 7, 11 and 13.

Multiples

A **multiple** is a number made by **multiplying together two other numbers**.

So the multiples of 2 are 2, 4, 6, 8, 10... and so on; of 3 are 3, 6, 9, 12, 15... and so on; of 10 are 10, 20, 30, 40, 50... and so on.

 Multiples of a number do not end at ten times the number, they go on and on. So, for example, 165 is a multiple of 5.

Venn diagrams

Test questions about factors and multiples often include **Venn diagrams**. These are **diagrams for sorting numbers**.
The numbers to 15 are sorted on this Venn diagram:

These numbers are factors of 30 and multiples of 3.

These numbers are neither factors of 30 nor multiples of 3.

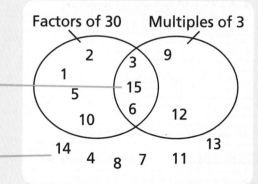

Rules of divisibility

This is a very complicated way of saying something very simple – these are rules to test whether a number is a multiple of 2, 3, 4, 5, 6, 8, 9 and 10.

Learn and use these rules of divisibility. A whole number is divisible by:

2 if the last digit is even.
Examples: 36, 194, 2116

3 if the sum of its digits can be divided by 3.
Examples: 135 (1 + 3 + 5 = 9)
2190 (2 + 1 + 9 + 0 = 12)

4 if the last two digits can be divided by 4.
Examples: 124, 340, 2564

5 if the last digit is 0 or 5.
Examples: 320, 145, 3025

6 if it is even and the sum of its digits is divisible by 3.
Examples: 528 (5 + 2 + 8 = 15)
402 (4 + 0 + 2 = 6)

8 if half of it is divisible by 4.
Examples: 264 (÷ 2 = 132)
432 (÷ 2 = 216)

9 if the sum of its digits is divisible by 9.
Examples: 135 (1 + 3 + 5 = 9)
207 (2 + 0 + 7 = 9)

10 if the last digit is 0.
Examples: 290, 350, 680, 410

Do you notice 7 is not there? That is because multiples of 7 have not really got a rule!

I'm glad Sam isn't a multiple! I couldn't bear him going on and on!

Have a go...

Explain to someone:

- *the factors of 24*
- *how you know a number is a multiple of 5*
- *the rule for finding numbers that can be divided by 3, 6 or 9.*

Key words

factor   multiple
prime number   Venn diagram

Quick Test

1. What are the six factors of 20?
2. Is 54 a multiple of 3?
3. Which two multiples of 4 are between 90 and 100?
4. True or false? 552 is divisible by 6.
5. Which number between 30 and 50 is a multiple of both 3 and 7?
6. What is the next prime number after 17?

Fractions

What is a fraction?

A fraction is **part of a whole number**.

¾ means 3 out of 4 equal parts.

There are three types of fractions:

- A **proper fraction**, such as $\frac{2}{5}$, which is less than 1.
- An **improper fraction**, such as $\frac{9}{4}$, which is greater than 1.
- A **mixed number**, such as $2\frac{1}{3}$, which is a whole number plus a proper fraction!

Parts of a fraction

A fraction has two parts:

$\frac{2}{3}$ ← numerator (top number)
← denominator (bottom number)

1 part out of 8 shaded. This shows $\frac{1}{8}$.

7 parts out of 8 shaded. This shows $\frac{7}{8}$.

If a chocolate fudge cake is divided into 5 equal pieces and I take one piece, what fraction is left?

$\frac{4}{5}$ of the cake is left. Not for long though...!

 Top Tip The denominator shows the number of equal parts. The numerator shows how many of them are taken. Remember that $\frac{2}{2}, \frac{3}{3}, \frac{4}{4}, \frac{5}{5}$... are all the same as one whole.

That pizza looks great – I definitely want more than you!

Typical! I know – you can have $\frac{3}{15}$ and I'll have the rest.

Equivalent fractions

Equivalent fractions are worth the same, even though they may look different.

Imagine eating $\frac{2}{4}$ (two-quarters) of a pizza – it is the same as eating $\frac{1}{2}$ (half) of the pizza.

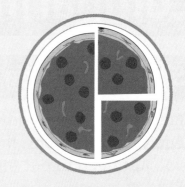

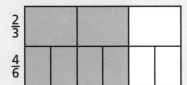

$\frac{2}{3} = \frac{4}{6} = \frac{6}{9} = \frac{8}{12}$

Fractions can be changed into their equivalent by either multiplying or dividing the numerator and denominator by the same amount.

Let me show you how:

$\frac{3}{7} = \frac{?}{21}$ $\qquad$ $\frac{3}{7} \xrightarrow{\times 3} \frac{9}{21}$ (×3)

Use this **equivalence wall** to find fractions that are worth the same.

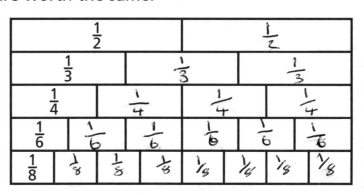

$\frac{15}{20} = \frac{3}{?}$ $\qquad$ $\frac{15}{20} \xrightarrow{\div 5} \frac{3}{4}$ (÷5)

Have a go...

Make a set of digit cards, 1 to 9. Shuffle them and take out any two cards.

- Make a fraction with the two digits.
- List four equivalent fractions to the one you have made.

Key words

numerator   equivalent fraction
denominator

Quick Test

1 What fraction is shaded? $\frac{7}{8}$

2 Which of these fractions is equivalent to $\frac{1}{3}$?

$\frac{3}{6}$ $\quad$ $\frac{4}{6}$ $\quad$ $\frac{2}{6}$

3 What fraction is equivalent to $\frac{10}{15}$?

4 Complete these:

a $\frac{3}{5} = \frac{12}{}$

b $\frac{18}{24} = \frac{}{4}$

c $\frac{5}{6} = \frac{25}{}$

Comparing fractions

Simple ordering

Putting fractions in order of size is a piece of cake if you know about **equivalent fractions**. If a set of fractions all have the same **denominator**, then you just need to put the **numerators** in order.

Example:

Put these in order, starting with the smallest: $\frac{3}{8}, \frac{7}{8}, \frac{5}{8}, \frac{1}{8}$

The correct answer is: $\frac{1}{8}, \frac{3}{8}, \frac{5}{8}, \frac{7}{8}$

Simple... **if all the fractions have the same denominator**.

Example:

Look at this example with mixed numbers:

Put these in order, starting with the largest: $1\frac{2}{5}, 3\frac{1}{5}, 1\frac{3}{5}, 2\frac{4}{5}$

Make sure you **order the whole numbers first** and then, if any whole numbers are the same, order the fraction parts.

$3\frac{1}{5}, 2\frac{4}{5}, 1\frac{3}{5}, 1\frac{2}{5}$

Tricky fractions

If a set of fractions have different denominators, then you need to **change them so that the denominators are all the same**.

Example:

Put these in order, starting with the smallest:

$\frac{3}{5} \quad \frac{7}{10} \quad \frac{1}{2} \quad \frac{1}{5}$

To make this easier, you need to make all the denominators the same by making **equivalent fractions**. To do this, look at the denominators and find a number that they can all divide into (**factors** of that number). They all divide into 10, or are factors of 10, so change them all to tenths:

$\frac{3}{5} = \frac{6}{10} \quad \frac{7}{10} = \frac{7}{10} \quad \frac{1}{2} = \frac{5}{10} \quad \frac{1}{5} = \frac{2}{10}$

Ordering is simple... if all the fractions have the same denominator: $\frac{2}{10}, \frac{5}{10}, \frac{6}{10}, \frac{7}{10}$

Important last step! **Remember to write out the fractions you started with:** $\frac{1}{5}, \frac{1}{2}, \frac{3}{5}, \frac{7}{10}$

Making the denominators the same is a good way of seeing which fraction is bigger than another.

There are two other methods: **using equivalence strips; and changing fractions to decimals.**

Using equivalence strips

If you need to work out which fraction is bigger, $\frac{2}{3}$, or $\frac{3}{4}$, you could draw equivalence strips for each.

Make sure the strips are the same length, then divide them into the number of denominators, and shade in the number of numerators.

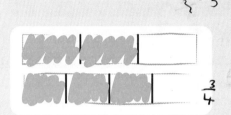

As you can see, $\frac{3}{4}$ is bigger than $\frac{2}{3}$.

Fractions to decimals

If you can change a fraction easily to a decimal, then you can put the decimals in order. **Remember to write the fractions in order at the end**. If you need to know more about decimals, go to page 22.

Example:

Put these in order, starting with the largest: $\frac{3}{5}$, $\frac{7}{10}$, $\frac{3}{4}$, $\frac{1}{2}$

$\frac{3}{5} = 0.6$ $\frac{7}{10} = 0.7$ $\frac{3}{4} = 0.75$ $\frac{1}{2} = 0.5$

So the order is 0.75, 0.7, 0.6, 0.5.

Write them out as fractions: $\frac{3}{4}$, $\frac{7}{10}$, $\frac{3}{5}$, $\frac{1}{2}$

Have a go...

Use a set of digit cards, 1 to 8, and put them into pairs to make four proper fractions.

- *Put them in order of size, starting with the smallest.*
- *Mix them up and repeat the activity.*

*If you are allowed to use a calculator for your test, it is simple to change a fraction to a decimal. Just see the line as a 'divide by' sign and type it in.
So $\frac{5}{8} = 5 \div 8 = 0.625$*

Key words

equivalent fraction   numerator
denominator   factor

Quick Test

1. Which is the larger fraction: $\frac{3}{5}$ or $\frac{5}{8}$?
2. Put these in order, starting with the smallest: $\frac{7}{10}$, $\frac{1}{10}$, $\frac{3}{10}$, $\frac{9}{10}$
3. Put these in order, starting with the smallest: $\frac{2}{3}$, $\frac{5}{6}$, $\frac{1}{3}$, $\frac{5}{9}$, $\frac{1}{2}$
4. What is $\frac{4}{5}$ as a decimal?
5. Which is the smaller fraction: $\frac{3}{10}$ or $\frac{5}{20}$?

Decimals and fractions

What are decimals?

Not all numbers are clear-cut whole numbers. **Decimals are 'in-between' numbers**, and are sometimes called decimal fractions. Look at these number lines.

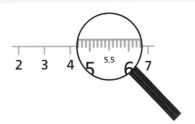

This shows tenths.

0.1 is the same as $\frac{1}{10}$

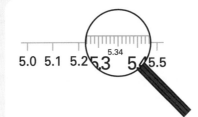

This shows hundredths.

0.01 is the same as $\frac{1}{100}$

Decimal points

A decimal point is used to separate whole numbers from decimals.

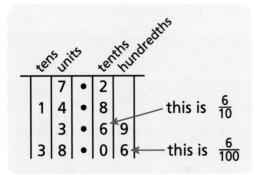

Top Tip: Remember that tenths are ten times larger than hundredths.

Fractions and decimals

Try to remember these fractions and decimals.

$\frac{1}{2}$ = 0.5 $\frac{1}{4}$ = 0.25 $\frac{3}{4}$ = 0.75 $\frac{1}{8}$ = 0.125

$\frac{1}{10}$ = 0.1 $\frac{1}{100}$ = 0.01 $\frac{3}{10}$ = 0.3 $\frac{1}{5}$ = 0.2

These two are recurring; they go on and on...

$\frac{1}{3}$ = 0.333... $\frac{2}{3}$ = 0.666...

× and ÷ by 10 and 100

Making a decimal ten or one hundred times bigger or smaller is easy, as long as you remember these rules:

To multiply by 10:

Move the digits **one place to the left** and fill the space with a zero

×10

4 . 3 1
4 3 . 1 0

To multiply by 100:

Move the digits **two places to the left** and fill the spaces with zeros

×100

0 . 6 5
6 5 . 0 0

To divide by 10:

Move the digits **one place to the right**

÷10

8 . 6
0 . 8 6

To divide by 100:

Move the digits **two places to the right**

÷100

1 4 . 3
0 . 1 4 3

Putting a zero on the end of a decimal does not change the number.

1.2 is the same as 1.20 and 1.200.

Have a go...

Using a calculator:

- *key in a fraction. For example, $\frac{3}{4}$ is 3 ÷ 4.*
- *look at the decimal answer.*

Which fractions make decimals that fill your calculator screen?

Key words

decimals

Quick Test

1. What is $\frac{7}{10}$ as a decimal?
2. Which fraction is the same as 0.4?
3. What is 2.86 × 10?
4. What is $\frac{3}{100}$ as a decimal?
5. What is 27.85 ÷ 100?

Comparing decimals

Ordering decimals

Putting decimals in order is just like putting whole numbers in order – **you need to look carefully at the value of each digit**.

Use these four easy steps to order decimals:

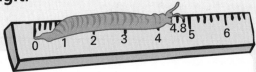

Example:

These are the lengths of six slugs you might find in your garden:

5.6 cm   4.85 cm   5.16 cm   5.08 cm   4.8 cm   4.78 cm

Write them in order of size, starting with the smallest.

1. Write them out, one under the other, lining up the decimal points.

   5.6
   4.85
   5.16
   5.08
   4.8
   4.78

2. Compare the digits from the left-hand side to the right.

   5.6
   4.85
   5.16
   4.78
   5.08
   4.8

3. For each digit, compare and write the smaller number. If they are the same, compare the next digit.

   4.78, 4.8, 4.85

4. Keep going until all the numbers are written out.

   4.78, 4.8, 4.85, 5.08, 5.16, 5.6

   If it helps, you could put a zero on the end of 5.6 and 4.8 so that all the decimals are the same length.

Comparing decimals

You may be asked to compare two decimals and say which is larger.

Example:
What is the missing sign?

< or > 2.34 ☐ 2.4

All you need to do is follow the four easy steps above to find the smallest decimal:

2.34 < 2.4

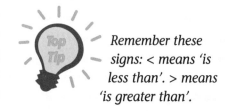

Remember these signs: < means 'is less than'. > means 'is greater than'.

Rounding decimals

Rounding decimals makes them easier to work with. For example, this caterpillar is 3.872 cm long.

This is very exact and you probably only need to know the caterpillar is about 4 cm long. If you want to be a little more accurate, you could say it is 3.9 cm long.

Decimals are rounded to the nearest whole number, the nearest tenth (1 decimal place) or nearest hundredth (2 decimal places).

Rounding to the nearest whole number:

Look at the tenths digit.

If it is 5 or more, round up to the next whole number.

If it is less than 5, the units digit stays the same.

8.5 rounds up to 9

3.46 rounds down to 3

Rounding to the nearest tenth, or rounding to 1 decimal place:

Look at the hundredths digit.

If it is 5 or more, round up to the next tenth.

If it is less than 5, the tenths digit stays the same.

6.76 rounds up to 6.8

5.347 rounds down to 5.3

Drawing a number line might help you with your rounding: for example, 3.42 is nearer to 3 than 4, so it rounds to 3.

 Have a go...

Use a calculator. Put in any number and divide it so you get a decimal answer.

- *Round it to the nearest whole number.*
- *Round it to one decimal place.*
- *Round it to two decimal places.*

 Key words

rounding

Quick Test

1. Which is the larger number, 4.15 or 4.7?
2. Place these decimals in order, starting with the smallest: 6.8, 6.12, 6.95, 6.08, 6.3
3. What is 12.53 to the nearest whole number?
4. Round 18.571 to the nearest tenth.
5. Round 6.384 to 2 decimal places.

Ratio and proportion

Proportion

Look at this **proportion** problem.

Patches is a brown cat that can be recognised by the white and ginger spots on her face. What proportion of her spots are white?

When you look at the proportion of an amount, it is the same as finding the **fraction of the whole amount**. So for this problem, what fraction of the spots are white?

There are 8 spots altogether. 2 of them are white, so $\frac{2}{8}$ of the spots are white.

This means that the proportion of white spots is 1 in every 4, or $\frac{1}{4}$.

Top Tip: *Remember that proportion means the fraction of the whole amount.*

Proportion patterns

Look at these tile patterns. What proportion of each of the patterns is blue?

Remember to find the fraction of the whole amount. Find the total number of tiles in each pattern and then find the fraction that is blue.

Ratio

Ratio is a little different to proportion because **it compares one amount with another**. Look at this family problem:

The favourite food of the Jones family is Yorkshire pudding and roast potatoes. They always have 1 Yorkshire pudding to every 3 potatoes. If the Jones family have 12 potatoes, how many Yorkshire puddings do they have?

The ratio of puddings to potatoes is 1 to 3, written as 1:3.

This ratio stays the same for different amounts:

Puddings	Potatoes
1	3
2	6
3	9

The ratio is 1:3 for all these, so if they have 12 roast potatoes, they would have 4 Yorkshire puddings.

Look at this example and follow the easy way to answering it.

Your friends mix 1 tin of yellow paint with every 2 tins of red paint. They need 12 tins of paint altogether. How many tins of red paint will they need?

With ratio problems it is a good idea to write the numbers in a chart:

yellow	red	total tins of paint
1	2	3
2	4	6
3	6	9
4	8	12

Your friends need 12 tins of paint, 8 of which are red.

Have a go...

Look in a cookery book and find some recipes with enough ingredients for four people.

- Change the amounts to feed 6, 8 or 10 people.
- Work out the ratio and proportions for different ingredients in a recipe.

Key words

proportion   ratio

Quick Test

1. If you use 4 tomatoes for every 1 litre of sauce, how many tomatoes will you need for 3 litres of sauce?

2. What proportion of these stars are red?

3. In a cricket club, there are 3 boys for every 2 girls. There are 25 children altogether. How many girls are there?

4. When making a pie, sugar and flour are mixed in the ratio of 1:3. If 320 g of sugar is used, how much flour is needed?

Percentages

What are percentages?

Percentages are simply **fractions out of 100** – that is what per cent means: out of 100. % is the percentage sign.

This pattern shows 25 out of 100 squares coloured red.

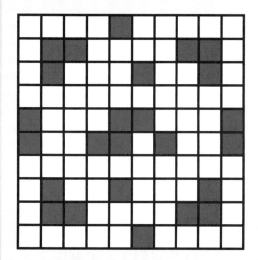

25% of the grid is red.

$\frac{25}{100}$ is equivalent to 25%

This pattern shows 5 out of 20 squares coloured red.

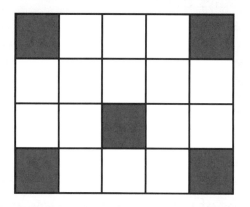

This is still 25% – but how do we know?

$\frac{5}{20} = \frac{25}{100} = 25\%$

If you need to remind yourself about working out equivalent fractions, look at page 20.

Top Tip — *Remember, numbers to be changed to percentages must be out of 100.*

Equivalence

It is a good idea to memorise these. Cover up different boxes and work them out.

Decimals	0.1	0.2	0.3	0.4	0.5	0.6	0.7	0.8	0.9	0.25
Fractions	$\frac{1}{10}$	$\frac{1}{5}$	$\frac{3}{10}$	$\frac{2}{5}$	$\frac{1}{2}$	$\frac{3}{5}$	$\frac{7}{10}$	$\frac{4}{5}$	$\frac{9}{10}$	$\frac{1}{4}$
Percentages	10%	20%	30%	40%	50%	60%	70%	80%	90%	25%

Converting percentages

Converting between percentages and decimals is easy:

per cent to decimal

Divide the percentage by 100.

Example

60% is the same as 0.6

decimal to per cent

Multiply the decimal by 100.

Example

0.25 is the same as 25%

If you are asked to convert a percentage to a decimal, remember that **it will always be less than 1**.

Converting between percentages and fractions is OK if you understand equivalent fractions:

per cent to fraction

Write the percentage as a fraction out of 100 and then simplify.

Example

40% is $\frac{40}{100}$ which is the same as $\frac{2}{5}$

fraction to per cent

Write the fraction as a decimal and then multiply by 100.

Example

$\frac{3}{4}$ is 0.75 which is the same as 75%

Have a go...

Look for percentages in newspapers and magazines.

- *Convert them to show them as fractions and decimals.*
- *Try to work out the information.*

Key words

percentage

Quick Test

1. What is 15% as a decimal?
2. What is 5% as a fraction?
3. Luke scored $\frac{16}{20}$ in a test. What percentage did he score?
4. What is $\frac{3}{5}$ as a percentage?
5. What is 0.01 as a percentage?

Test your skills

Grid challenge 1

Write the numbers 1 to 16 on small pieces of paper.

Place the numbers on the grid so that they follow the rules.

For example, the top left-hand square must be an even number and a square number.

	even number	factor of 24	< 8	odd number
square number				
> 6				
multiple of 3				
< 12				

Can you place all the numbers on the grid?

Grid challenge 2

Now try the same with this grid.

	square number	even number	odd number	< 12
triangular number				
multiple of 3				
multiple of 2				
> 8				

Which are impossible squares?

How many squares can you complete?

Make your own grid challenge.

Can you place all the numbers on the grid?

Test your knowledge

Section 1

1. Write the number 27 384 in words.
   _____

2. What is the largest number you can make from these digits?
   1   0   6   8   2   _____

3. 64 800 ÷ 100 = ?   _____

4. Put these numbers in order, starting with the largest:
   300   1751   51 001   4308   756   6117
   _____ _____ _____ _____ _____ _____

5. Round 4503 to the nearest 1000.   _____

Section 2

1. Put these numbers in the correct places.
   489   240   420

   [420] > [＿＿] > [＿＿]

2. What is 5^2?   _____

3. $3s - 1 = 20$ What is the value of s?   _____

4. What are the factors of 30?   _____

5. Is 66 a multiple of 9?   _____

6. What is the square root of 49?   _____

7. True or false?
   615 is divisible by both 5 and 3.   _____

8. Round 3728 to the nearest 10.   _____

9. What are the missing numbers in this table?

IN	3	8		9	2	5
OUT	12		16	36		20	28

   FUNCTION

Section 3

1. Which of these fractions is equivalent to $\frac{1}{3}$? $\frac{2}{3}, \frac{3}{4}, \frac{2}{6}, \frac{2}{4}$ _____

2. Which is the larger fraction: $\frac{3}{4}$ or $\frac{5}{8}$? _____

3. What is $\frac{2}{5}$ as a decimal? _____

4. Which fraction is the same as 0.3? _____

5. What is 3.51 × 100? _____

6. What is 17.2 ÷ 10? _____

7. Place these lengths in order, starting with the shortest:

    4.05 m   2.3 m   4.23 m   3.44 m

    _____   _____   _____   _____

Section 4

1. David shares out 25 cards. He gives Callum one card for every four he keeps. How many cards does Callum get? _____

2. What proportion of the squares are red? _____

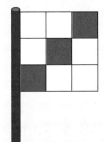

3. Ali got 15 out of 20 in a spelling test. What percentage did he score? _____

4. What are the missing decimals, fractions and percentages in this chart?

Decimals	0.5	0.1	0.3	0.4	0.25	
Fractions		$\frac{1}{10}$			$\frac{1}{4}$	$\frac{3}{5}$
Percentages			30%	40%		60%

Mental maths

Number facts

You need to know all your number bonds to 20. These are all the addition facts to 20, and from these you can learn the subtraction facts. For example, 7 + 8 = 15 and 15 – 8 = 7. It is very simple – learn these and then use them to help you with trickier sums.

3 + 9 = 12
9 + 3 = 12
12 – 3 = 9
12 – 9 = 3

+	0	1	2	3	4	5	6	7	8	9	10
0	0	1	2	3	4	5	6	7	8	9	10
1	1	2	3	4	5	6	7	8	9	10	11
2	2	3	4	5	6	7	8	9	10	11	12
3	3	4	5	6	7	8	9	10	11	12	13
4	4	5	6	7	8	9	10	11	12	13	14
5	5	6	7	8	9	10	11	12	13	14	15
6	6	7	8	9	10	11	12	13	14	15	16
7	7	8	9	10	11	12	13	14	15	16	17
8	8	9	10	11	12	13	14	15	16	17	18
9	9	10	11	12	13	14	15	16	17	18	19
10	10	11	12	13	14	15	16	17	18	19	20

Rounding

If you need to add or take away 9 or 19, round the number to 10 or 20 to make it easier.

Examples:

16 + 9
Add 10, take away 1
Answer : 25

37 + 19
Add 20, take away 1
Answer: 56

34 – 9
Take away 10, add 1
Answer: 25

63 – 19
Take away 20, add 1
Answer: 44

You can also use this method to add or take away larger numbers ending in 9 such as 29, 39 and 49.

Adding 2-digit numbers

If you know your number bonds, then adding bigger numbers in your head is easier.

Example:

35 + 57   Use these three steps:

1 **Hold the bigger number in your head**: 57

2 **Add the tens**: 57 + 30 = 87

3 **Add the units**: 87 + 5 = 92

Subtracting 2-digit numbers

A really good method for subtraction is to **find the difference between the numbers by counting on**.

Example:

93 – 56   This number line shows exactly what goes on in your head.

Count on from 56 to 60. Hold the 4 in your head.

60 to 93 is 33   33 + 4 is 37   So 93 – 56 = 37

 Top Tip *If it helps, draw a quick number line and show the steps. Remember to put the smaller number on the left and the larger number on the right.*

Have a go...

Shuffle your digit cards, 1 to 9, and take any four cards. Make two 2-digit numbers.

- *Find the total of the two numbers.*
- *Find the difference between the two numbers.*
- *Rearrange the four digits. What is the largest total and smallest difference you can make?*

Key words

difference

Quick Test

1. What is 45 + 46?
2. What is 74 – 19?
3. What is the sum of 64 and 87?
4. What is the difference between 83 and 58?
5. What is the total of 14, 15 and 16?
6. What is 74 subtract 36?

Multiplication facts

Times tables

×	0	1	2	3	4	5	6	7	8	9	10
0	0	0	0	0	0	0	0	0	0	0	0
1	0	1	2	3	4	5	6	7	8	9	10
2	0	2	4	6	8	10	12	14	16	18	20
3	0	3	6	9	12	15	18	21	24	(27)	30
4	0	4	8	12	16	20	24	28	32	36	40
5	0	5	10	15	20	25	30	35	40	45	50
6	0	6	12	18	24	30	36	42	48	54	60
7	0	7	14	21	28	35	42	49	56	63	70
8	0	8	16	24	32	40	48	56	64	72	80
9	0	9	18	27	36	45	54	63	72	81	90
10	0	10	20	30	40	50	60	70	80	90	100

I wonder how many times you have been told to learn your tables! Well, you are being told again – "Learn your times tables!"

$9 \times 3 = 27$

$3 \times 9 = 27$

$27 \div 9 = 3$

$27 \div 3 = 9$

Here are all the multiplication and division facts to 100. Cover up different numbers in the grid and say the missing numbers as quickly as possible.

Tricky tables

These are the facts that cause the most problems.

Learn one fact a day – it will only take 10 days! Try this: every time you go through a doorway at home, say the fact out loud. You will soon know it off by heart.

Division facts are also important to learn – they are just the **inverse**, or opposite, of multiplication facts: $6 \times 9 = 54 \quad 54 \div 6 = 9$

$3 \times 8 \qquad 6 \times 7$

$7 \times 9 \qquad 6 \times 8$

$4 \times 7 \qquad 9 \times 6$

$4 \times 8 \qquad 8 \times 9$

$4 \times 9 \qquad 7 \times 8$

Top Tip

Remember that 3×8 gives the same answer as 8×3, so you only have to learn half the facts.

Mental calculations

If you know your tables, then multiplying 2-digit numbers by a single digit should be easy.

Example: 63 × 4

Always use these three steps:

1 **Multiply the tens:** 60 × 4 = 240
2 **Multiply the units:** 3 × 4 = 12
3 **Add the two parts:** 240 + 12 = 252

To multiply tens by a single digit, work out the fact and then make it ten times bigger:

60 × 4 = 6 × 4 × 10 = 24 × 10 = 240

Missing number problems

Maths tests often have missing number questions.

With questions like this, always use the numbers you are given to work out the missing number.

Important things to remember are:

Multiplication and division are inverses (opposites).

Division can be checked by multiplication.

Examples:

6 × ☐14☐ = 84

84 ÷ 6 = 14

☐☐ ÷ 9 = 15

15 × 9 = 135

Have a go...

Get your friends to give you a tables speed test.

- *They write 30 tables facts down on paper to be read out.*
- *You write the answers.*
- *Keep a record of your time and score.*
- *Have another go and try to beat your best time and score.*

9 times table is my favourite – just add the digits and they always make nine.

Let me try. 8 × 9 is 72. 7 + 2 is 9. Clever!

Key words

inverse

Quick Test

1. Answer these as quickly as you can:

   a 4 × 9   b 6 × 8   c 7 × 4   d 8 × 7

2. What is 34 × 6?

3. What is the missing number in each of these?

   a ☐ × 4 = 92   c 6 × ☐13☐ = 78
   b ☐ ÷ 6 = 13   d 144 ÷ ☐12☐ = 12

Addition and subtraction

Written addition

If you are given a sum and the numbers are too big, or there are too many numbers to add in your head, then you need to use a written method.

Follow this step-by-step method:

Example:

3527 + 768

1 **Write the numbers neatly in a column, lining up the digits.**

```
  3527
+  768
──────
```

2 **Start by adding from the right-hand column**, the units column. For any total over 9 just put the tens digit under the next column.

```
  3527
+  768
──────
     5
     1
```

3 Now do the same with the tens column.
Keep going left until all the columns have been added.

```
  3527
+  768
──────
  4295
  1 1
```

Adding decimals

When you add decimals, remember to line up the decimal points. The method is the same as with whole numbers:

Example:

17.9 + 8.63

1 **Write them in a column, lining up the decimal points.**

```
  17.90
+  8.63
───────
  26.53
   1
```

2 **Start by adding from the right-hand column.**

```
  17.90
+  8.63
───────
  26.53
   1
```

3 **Keep going left until all the columns have been added.**

```
  17.9
+ 8.63
───────
 26.53
  11
```

Write the decimal point in the answer space before you start. It should line up with the decimal points above.

Written subtraction

Here are two good ways to subtract. Have a go at both methods and choose the one that you feel happier with – then practise, practise, practise!

Method 1:

1. **Write the numbers in a column, lining up the units digits.**

2. **Start from the right-hand column, take away the bottom number from the top number.**

3. **Now do the same with the other columns. If the top number is smaller than the bottom number**, then, in this example, **exchange** a 100 from the hundreds column. You are making the 40 into 140 and the 300 into 200.

Method 2: Find the **difference** between the numbers by counting on.

1. **Draw a blank number line.**

2. **Count on to the next ten, then to the next hundred, then to the number.**

3. **Add up all the jumps.**

   $246 + 10 + 6 = 262$

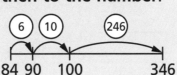
84   90 100   346

Have a go...

Explain to someone the method you use for:

- adding two large numbers
- subtracting two large numbers.

Important! Remember to always take the bottom number away from the top number. Write it nice and clearly.

Key words

difference

Quick Test

1. 345 + 267 = 612
2. 709 + 426 = 1185
3. 1485 + 2097 = 3592
4. 483 − 127 = 355
5. 506 − 329 = 177

Multiplication methods

Long multiplication

If you have a question like 68 × 53... do not panic!

It helps if you **know your tables** and then you just need to be organised.

Have a go at these two methods and choose the one you prefer... then practise and practise again!

If you want to practise your tables, turn back to page 36.

Top Tip: Make sure you keep all the columns lined up – do not squash up the numbers or you will get confused.

Column method

Example:
68 × 53

Before you start, **estimate** an **approximate** answer. 70 × 50 is 3500, so the answer will be close to 3500.

1. Write the numbers in a column, lining up the units digits.

   ```
       68
   ×   53
   _____
   ```

2. Start with the units.

   68 × 3 = 204

   8 × 3 = 24
   60 × 3 = 180

   ```
       68
   ×    3
   _____
      204
        2
   ```

3. Now do the same with the tens.

   68 × 50 = 3400

   8 × 50 = 400
   60 × 50 = 3000

   ```
       68
   ×   50
   _____
      400
   + 3000
   _____
     3400
   ```

4. Add up the two totals.

   ```
       68
   ×   53
   _____
      204
   + 3400
   _____
     3604
   ```

These look tricky - can't I just use a calculator?

You'd probably get in a muddle with the keys - if only you were as organised and careful as me!

Grid method

Example: 68 × 53

1. Write the numbers in a grid, breaking up each number into tens and units (TU).

×	50	3
60		
8		

2. Multiply each pair of numbers to complete the grid.

×	50	3
60	3000	180
8	400	24

3. Add up the rows in the grid.

×	50	3	
60	3000	180	3180
8	400	24	424

4. Add up the two totals.

×	50	3	
60	3000	180	3180
8	400	24	+ 424
			3604

Look at the column method and the grid method carefully and choose the one you prefer.

Once you understand it, **HTU** x TU can be worked out using the same methods:

```
   415
 ×  38
  3320
 12450
 15770
```

	400	10	5	
30	12000	300	150	12450
8	3200	80	40	3320
				15770

Have a go...

Explain to someone the written method you use for:

- multiplying two 2-digit numbers
- multiplying together a 2-digit and a 3-digit number.

Key words

estimate   HTU

approximate

Quick Test

Choose a written method to answer these:

1. 34 × 18 = 306
2. 64 × 75
3. 93 × 56
4. 192 × 43
5. 286 × 74

```
   34
 × 18
  272
+ 834
  306
```

Division methods

Division and multiplication

The most important thing to remember about division is that **it is the inverse (opposite) of multiplication**.

So if you know your tables it will really help you to divide numbers.

$8 \times 3 = 24 \qquad 24 \div 3 = 8 \qquad 24 \div 8 = 3$

There are several ways of writing a division:

$42 \div 6 = 7 \qquad \frac{42}{6} = 7 \qquad 6\overline{)42}\,^{7}$

All these mean 42 divided by 6.

If you want to practise your tables, go back to page 36.

Written method

Before you start, estimate an approximate answer, then you can get going on the calculation.

$258 \div 6$ is approximately $300 \div 6$, so the answer will be less than 50.

$$\begin{array}{r} 43 \\ 6\overline{)258} \\ 240 \\ \hline 18 \\ 18 \\ \hline 00 \end{array}$$

Step 1: How many sixes in 250? The answer must be a multiple of 10, so it is 40 sixes, or 240, with 10 remaining.

Step 2: Add the 10 to the 8 to show the remaining number to be divided.

Step 3: How many sixes in 18? 3 sixes are 18. Write the answer at the top, 43, and check with your estimate.

Once you understand this, you could try writing it using a shorter method:

$$6\overline{)25_{1}8}\,^{4\ 3}$$

Try to follow the three steps.

Remainders

Many division answers are not exact – they have an amount left over. For example, if you had 14 sweets and wanted to share them between 5 of you, you would have 2 sweets each and 4 left over. Imagine the arguments!

Round up: 68 people are going on a trip. 9 people can fit in each minibus. How many buses are needed?

68 ÷ 9 is 7 remainder 5, so 8 minibuses are needed.

Round down: I have £68. How many £9 CDs could I buy for that amount?

68 ÷ 9 is 7 remainder 5, so 7 CDs could be bought.

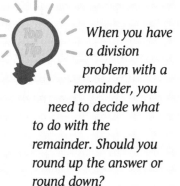

When you have a division problem with a remainder, you need to decide what to do with the remainder. Should you round up the answer or round down?

Written method

This is exactly the same as divisions without remainders – except there will be something left over.

187 ÷ 4 is approximately 200 ÷ 4, so the answer will be roughly 50.

Step 1: How many fours in 180? The answer must be a multiple of 10, so it is 40 fours, or 160, with 20 remaining.

Step 2: Add the 20 to the 7 to show the remaining number to be divided.

Step 3: How many fours in 27? 6 fours are 24.

Step 4: 3 is left over. Write the answer at the top, 46 remainder 3, and check with your estimate.

Have a go...

Choose four cards from your digit cards, 1 to 9, and practise:

- *dividing a 3-digit number by a single digit*
- *divisions with remainders.*

Key words

inverse

Quick Test

1. What is 68 divided by 4?
2. Calculate 483 ÷ 9.
3. What is £385 shared by 5 people?
4. A box holds 6 eggs. How many boxes will be needed for 205 eggs?
5. A factory makes 140 lamps in a day. 3 lamps are packed in each box. How many boxes are filled?

Fractions of quantities

Fractions and numbers

You are sure to get asked questions like this in a test:

> A squirrel stores 40 nuts in a tree hollow. A fox comes along and steals $\frac{3}{5}$ of the nuts. How many nuts does the squirrel have left?

You need to understand a lot about fractions to answer this problem, so take **one step at a time**.

Simple fractions

Look at these examples: What is: $\frac{1}{3}$ of 12, $\frac{1}{5}$ of 20 and $\frac{1}{4}$ of 24?

These fractions all have 1 as a **numerator**.

So simply divide by the denominator.

$\frac{1}{3}$ of 12 is $12 \div 3 = 4$ $\frac{1}{5}$ of 20 is $20 \div 5 = 4$ $\frac{1}{4}$ of 24 is $24 \div 4 = 6$

Tricky fractions

Look at these examples:

What is $\frac{2}{3}$ of 12, $\frac{4}{5}$ of 20 and $\frac{3}{4}$ of 24?

For these fractions, we know how to manage with 1 as a numerator.

If it is more than 1, divide by the denominator and then multiply by the numerator.

Read that sentence again so you understand.

$\frac{2}{3}$ of 12 is $12 \div 3 = 4$, then $\times 2 = 8$

$\frac{4}{5}$ of 20 is $20 \div 5 = 4$, then $\times 4 = 16$

$\frac{3}{4}$ of 24 is $24 \div 4 = 6$, then $\times 3 = 18$

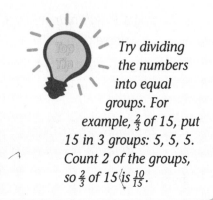

Try dividing the numbers into equal groups. For example, $\frac{2}{3}$ of 15, put 15 in 3 groups: 5, 5, 5. Count 2 of the groups, so $\frac{2}{3}$ of 15 is $\frac{10}{15}$.

Solving the problem

Read the squirrel and fox problem again.

You need to find $\frac{3}{5}$ of 40. $\frac{1}{5}$ of 40 is 40 ÷ 5 = 8.
So $\frac{3}{5}$ is 8 × 3 which is 24.

The fox takes 24 nuts, so 16 nuts are left for the squirrel.

 Top Tip *Remember that proportion means the fraction of the whole amount.*

Fractions of shapes

You can find fractions of shapes in the same way.

Example:

$\frac{3}{4}$ of this box of chocolate is for Mum and the rest is for Dad. How many chocolates will Dad get?

$\frac{1}{4}$ of 16 is 16 ÷ 4 = 4

"Poor squirrel, losing his nuts."

"Poor me, having to listen to you. You're nuts!"

Have a go...

Use a pile of different coins to make 60p. Sort the coins into three sets showing: $\frac{1}{5}$, $\frac{1}{2}$, $\frac{1}{4}$ of 60p.

Remember to divide by the denominator. For example, $\frac{1}{5}$ of 60 is $\frac{60}{5} = 12$.

So $\frac{1}{5}$ of 60p is 12p.

Now try the others.

How much money have you got left over each time?

Try this with other amounts.

Key words

numerator   denominator

Quick Test

1. What is $\frac{1}{4}$ of 8?
2. What is $\frac{5}{6}$ of 24?
3. Jake has £2 pocket money. If he spends $\frac{3}{8}$ of his money on sweets, how much does he have left?
4. A man weighs 90 kg. His daughter is $\frac{2}{3}$ of his weight. How much heavier is the man than his daughter?
5. A football has 32 red and white hexagonal panels. If $\frac{3}{4}$ of them are red, how many are white?

Percentages and money

Percentages of a quantity

What is 20% of £60?

This is a typical **percentage** question and there are several ways to solve it. Whichever way you choose, remember that the **word 'of' means multiply**.

Method 1:

Change to a fraction and work it out:

$\frac{20}{100} \times £60 = 20 \times \frac{20 \times 60}{100} = \frac{1200}{100} = £12$

Method 2:

Use 10% to work it out. Just divide by 10:

10% of £60 is £6. So, 20% of £60 is double that: £12

Method 3:

If you are allowed, **use a calculator** to work it out: Key in 20 ÷ 100 x 60 = £12

Try each of these methods to work out:

30% of 50   25% of 80   60% of 90   5% of 60

Which is your favourite method?

To find 5%, remember that it is half of 10%. To find 20%, remember that it is double 10%.

What would you rather have: 20% of £80, or 80% of £20?

They both sound a lot more than I've got at the moment!

Discounts and sale prices

If you know how to work out percentages of amounts, you can work out sale prices. There are always two steps to remember:

Example:

A fancy dress shop has a 20% sale. If a gorilla suit costs £32, what is the sale price?

Use these two steps to work out the sale price of a set of chattering teeth costing £6.

Step 1
Work out the percentage: Work out 10% and double it.
20% of £32 is £6.40.

Step 2
Take away this amount from the price:
32 – 6.40 is 25.60.

So the sale price is £25.60.

Percentage increases

If prices increase, there are still two steps, but you add the percentage to the price instead of taking it away:

Example:

The price of a £340 violin is increased by 5%. What is the new price?

Step 1: Work out the percentage: 5% of £340 is £17.

Step 2: Add this amount to the price: 340 + 17 is 357.

So the new price of the violin is £357.

Have a go...

Look at the price labels of different items around the house or shops.

- Discount the prices by reducing them by 10%.
- Make up a sales poster with your information.
- Change the percentage to 15%, 50%, 80%... or whatever you fancy!

Key words

percentage

Quick Test

1. What is 40% of £15?
2. What is 15% of £200?
3. A hat costing £10 is reduced in a sale by 25%. What is the sale price?
4. Prices in a baby shop are increased by 10%. What is the new price of a pack of nappies that originally cost £8.40?

Problems

Word problems

Many test questions are disguised as tricky **word problems**. When you see one, do not panic and follow these four steps.

Example:
Four tickets for the circus cost £42. What is the cost of three tickets?

Step 1: **Read the problem.** Try to picture the problem and imagine going through it in real life.

Step 2: **Sort out the calculations.** 42 ÷ 4 will give the price of one ticket. Then multiply the answer by 3 to find the cost of three tickets.

Step 3: **Answer the calculations.**

42 ÷ 4 = 10.5

10.5 × 3 = 31.5

Step 4: **Answer the problem.** Look back at the question – what is it asking?

The cost of three tickets is £31.50.

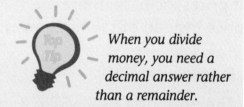

When you divide money, you need a decimal answer rather than a remainder.

Answering problems

Word problems can have different numbers of calculations to work out before you reach the final answer. Follow the four steps above and answer these:

One-stage problems (one calculation)

Ryan scored 3845 points on a computer game. The highest score is 4906. How many points is he behind the highest score?

Two-stage problems (two calculations)

Natalie wants to buy four books that cost £3.20 each. She has £10. How much more money does she need?

Three-stage problems (three calculations)

Ali has a packet of 24 biscuits. He ate ⅓ of them on Monday and six more on Tuesday. How many biscuits are left?

Using a calculator

Make sure you know how to use these keys on a calculator.

+/− : changes a **positive number** to a **negative number**

% : percentage, e.g. to find 40% of £60, key in 60 × 40% is £24. Don't press =

√ : **square root** (inverse of a **square number**), e.g. to find square root of 144 press 1 4 4 then the √ key

C or CE : clears the last entry, useful if you make a mistake halfway through a sum

AC : clears all entries and leaves 0

Many calculators have different keys, so get used to the one you have and practise different calculations.

When you can use a calculator to solve money problems, you need to remember:

A display of 3.8 means £3.80

A display of 2.06 means £2.06

A display of 0.54 means £0.54 or 54p

A display of 3.48719 needs to be rounded to 2 decimal places: £3.49

Have a go...

Explain to someone:
- *the four steps to solving word problems*
- *the different keys on a calculator.*

Key words

positive number   square root
negative number   square number

Quick Test

You can use a calculator with these questions.

1. I think of a number, then divide it by 12. The answer is 30. What was my number?

2. In a gym display there are 3 girls and 2 boys in each group. 42 girls take part in the display. How many boys are there altogether?

3. 268 children and 27 adults went on a school trip. Buses seat 64 people. How many buses were needed?

4. Harry buys 4 comics that cost 90p each. How much change from £5 will he get?

Test your skills

Digit sums

Use the digits 1, 2, 3 and 4.

Use the ×, + and − signs, and brackets.

Make sums to give the answers 1 to 30.

Examples:

1. (2 × 3) − (4 + 1) = 1
2. (2 + 3) − (4 − 1) = 2

Examples:

1. 12 − (3 + 4) = 5
2. (4 − 2) × 13 = 26

Try to use all the digits in each sum

Remember, the digits can make larger numbers, e.g. 12

Which numbers can be made in the most ways?

Digit switch

Use a calculator to investigate sums like this:

$$12 \times 84 = 1008 \xrightarrow{\text{switch digits}} 21 \times 48 = 1008$$

$$42 \times 36 = 1512 \xrightarrow{\text{switch digits}} 24 \times 63 = 1512$$

Find other pairs of 2-digit numbers which do this.
You can use numbers from 10–99.

What do you notice about the numbers that can be reversed?

Which numbers do not work?

I wish I could do a sister switch.

His days as my brother are numbered!

Test your knowledge

Section 1

1. 45 + 37 = ?

2. What is the sum of 89 and 33?

3. What is the difference between 17 and 56?

4. What is 82 − 29?

5. 54 × 7 = ?

6. 3 × ☐ = 81

7. What is 57 multiplied by 3?

8. Share £192 between 8 people.

   You can use a written method to help you.

Section 2

1. a 1520 + 3769   _5289_
   b 841 − 503   _328_
   c 2295 − 467   _1828_

2. a 31 × 59
   b 702 ÷ 9
   c 361 ÷ 6

Section 3

1. Six CDs cost £45. What is the cost for 5 CDs?

2. What is £273 shared by 3 people?

3. My Gran is twice as old as my Mum, who is 3 times older than me. If Gran is 72 years old, how old am I?

4. A jug holds 174 ml of juice. It is poured equally between 6 glasses. How much juice is there in each glass?

Section 4

You can use a calculator with the next 6 problems.

1. I think of a number, then multiply it by 6 and the answer is 132. What was my number? _____

2. Imran reads 26 pages of his book each day. He started the book on Monday. After reading on Friday he still had 182 pages to read. How many pages are there in total in the book? _____

3. A pet shop has 56 rabbits and 38 guinea pigs. Every day each rabbit eats 25 g and each guinea pig eats 15 g of food. How much food is needed in total for one day? _____

4. Jan buys four bananas costing 39p each. How much change will she get from £2? _____

5. Look at the price list on the right. How much is saved if 2 adults and 2 children buy a family ticket?

   _____

   NOW OPEN — CITY POOL
   FEES:
   Adult swim £2.25
   Child swim £1.75
   Family swim £7.00
   (2 adults and up to 3 children)

6. Would it be cheaper for 1 adult and 3 children to buy a family ticket? _____

Section 5

1. Ali got 15 out of 20 in a spelling test. What percentage did he score? _____

2. What is 75% of 60? _____

3. A CD player costs £70. It is reduced in a sale by 20%. What is the sale price? _____

4. A normal-size chocolate bar weighs 60 g and a king-size bar has 20% extra. How much does the king-size bar weigh? _____

5. There are 24 eggs in a box. $\frac{3}{8}$ of the eggs are broken. How many eggs are not broken? _____

2-D shapes

Polygons

Any shape with **straight sides** is called a **polygon**. The name of the polygon tells you the number of sides:

Number of sides	Name
3	Triangle
4	Quadrilateral
5	Pentagon
6	Hexagon
7	Heptagon
8	Octagon

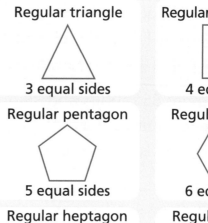

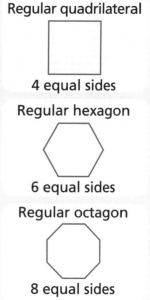

Triangles

Try to remember these triangles…

Equilateral	Isosceles	Right-angled	Scalene

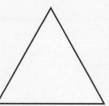  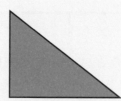 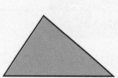

- 3 equal sides
- 3 equal angles

- 2 equal sides
- 2 equal angles

- 1 angle is a right angle (90°)

- no equal sides
- no equal angles

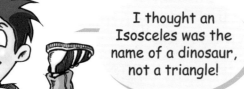

I thought an Isosceles was the name of a dinosaur, not a triangle!

Maybe you're thinking of a Triceratops – it had three horns!

Quadrilaterals and circles

Learn about these special four-sided shapes:

Square

- 4 equal sides
- 4 right angles

Parallelogram
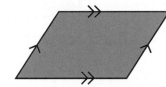
- opposite sides are equal and parallel

Circle
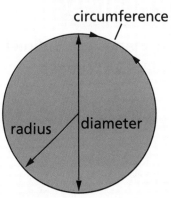

The **diameter** is the **distance right across the circle** through the centre.

The **radius** is the **distance from the centre of the circle to the edge**.

The **circumference** is the **distance all the way around**.

Rectangle
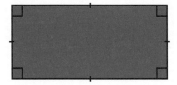
- 2 pairs of equal sides
- 4 right angles

Kite

- 2 pairs of **adjacent** sides are equal

Rhombus
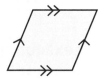
- 4 equal sides
- opposite angles equal
- opposite sides **parallel**

Trapezium
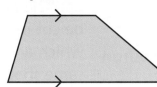
- 1 pair of parallel sides

 Top Tip *A square is a special rectangle and a rectangle is a special parallelogram. Therefore, a square is actually a type of parallelogram.*

Have a go...

- *Write down the names of some of the shapes you have just read about.*
- *Now close the book and sketch the shapes.*
- *Describe all the properties.*

Key words

polygon	diameter
parallel	radius
adjacent	circumference

Quick Test

1. What is the name of a five-sided polygon?
2. What is special about regular polygons?
3. Which triangle has 2 equal sides and 2 equal angles?
4. If the radius of a circle is 12 cm, what is its diameter?
5. What is another name for a regular triangle?

3-D solids

3-D shapes

Solid shapes are **3-dimensional**. You need to learn the names and properties of these 3-D shapes.

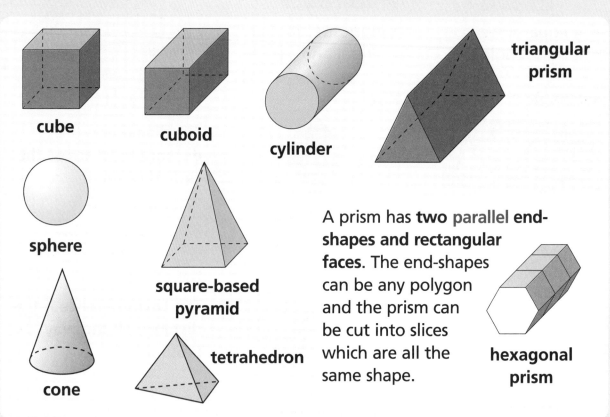

A prism has **two parallel end-shapes and rectangular faces**. The end-shapes can be any polygon and the prism can be cut into slices which are all the same shape.

Activity:

- Look at one shape and remember its name.
- Close your eyes and picture the shape floating on the back of your eyelids.
- Turn the shape around and try to picture it from different angles.
- Try this with other shapes so that you really get to know them.

I'm picturing a shape. I imagine holding it carefully with the point downwards and two scoops of strawberry ice-cream at the top...

Typical – always thinking of food!

Parts of a 3-D shape

3-D shapes are made up of **faces**, **edges** and **vertices** (corners).

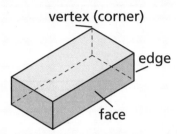

A cuboid has 6 faces, 12 edges and 8 vertices.

A face is a flat surface of a solid.

An edge is where two faces meet.

Vertex is another word for corner. The plural is vertices.

Nets of solids

The **net** of a shape is what it looks like when it is opened out flat. If you carefully pull open a cereal box so that it is one large piece of cardboard, you will have the net of the box. I suggest you do this after you finish the cereal to avoid a mess!

Examples

Net of a triangular prism.

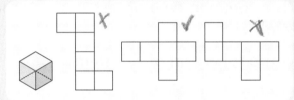

Some nets of a cube.

Top Tip: *Most shapes have more than one net, like the cube. A popular test question is to spot the odd one out in a set of nets of cubes. One of them is not a proper net. Try to imagine folding each net up from one of the centre squares, and you should be able to spot the one that does not make a cube.*

Have a go...

Find some 3-D shapes from packets and boxes around the house.

- Sketch the shapes.
- Name them.
- List the number of faces, edges and vertices.

Key words

- 3-dimensional
- parallel
- face
- edge
- vertex
- net

Quick Test

1. Which shape has 4 triangular faces and a square face?
2. What is another word for the vertex of a shape?
3. How many edges does a tetrahedron have?
4. What is the name of the shape made from this net?
5. How many faces does a hexagonal prism have?

Geometry

Lines of symmetry

A shape is **symmetrical** if **both sides match when a mirror line is drawn**. This is sometimes called **a line of symmetry**. Some shapes have more than one line of symmetry:

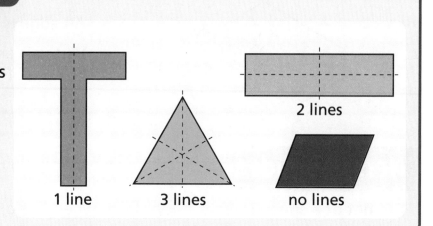

1 line   3 lines   2 lines   no lines

Reflections

This is a common type of test question:

Draw the reflection of this shape.

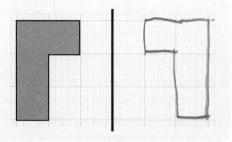

Imagine the line is a mirror. Draw dots on each corner and count the squares across so that each point is reflected:

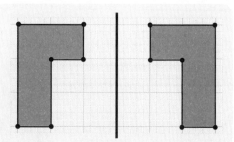

Rotational symmetry

If you can turn or rotate a shape and it still looks the same, then it has rotational symmetry. The order of rotational symmetry is the number of times the shape can turn and still look the same.

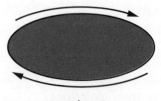 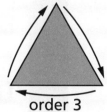

order 2   order 3   order 4

Moving shapes

A shape can be moved by:

- Translation: this is a complicated word for a simple idea – sliding a shape without rotating or flipping over.

This shape has moved 4 squares across and 1 square down.

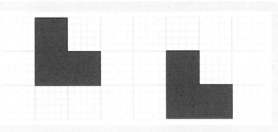

- Reflection: this is sometimes called flipping over.

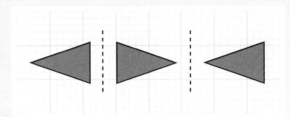

- Rotation: a shape can be rotated about a point, **clockwise** or **anti-clockwise**.

Shape A is rotated around point X to become shape B.

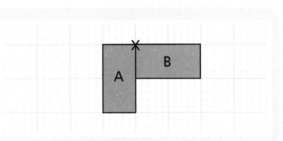

Have a go...

Draw some shapes. Show someone shapes with:

- lines of symmetry
- rotational symmetry.

Draw more shapes and patterns to show:

- rotation
- reflection
- translation.

Key words

symmetrical   anti-clockwise
clockwise

On test questions about symmetry you are allowed to use a small mirror to draw reflected shapes. Practise doing this – it is a useful way of checking if your shape is reflected correctly.

Quick Test

1. How many lines of symmetry has an isosceles triangle?
2. How many lines of symmetry has this shape?
3. A parallelogram has rotational symmetry of order 2. True or false?
4. Has this shape been reflected or rotated?

Using coordinates

Finding positions

Coordinates are used to **show an exact position of a point on a grid**. Two numbers from the x- and y-axes show the position:

Coordinates are always written in brackets and separated by a comma.

The coordinates of A are (3,4)

The coordinates of B are (5,2)

Top Tip: *The numbers on the horizontal x-axis are written first, then the y-axis. You can remember this because x comes before y.*

Position (0,0) is also called the origin.

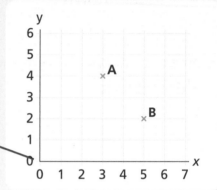

Making shapes

You may have a test question a bit like this: Here are three corners of a rectangle.

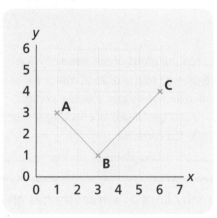

- What are the coordinates of these three corners?
- What would be the coordinates of a fourth corner, D?

Remember to **read the numbers across and then up for each position**.

Lightly draw in the missing lines. Use a ruler and be as accurate as you can.

I'm going to plot the coordinates of the fridge!

Typical!

Negative numbers

Coordinates also use **negative numbers** to show position.

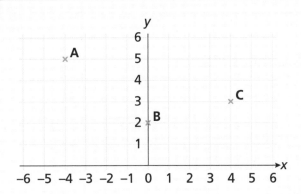

The coordinates of A are (–4,5).

The coordinates of B are (0,2).

The coordinates of C are (4,3).

This grid has 4 **quadrants** with negative numbers on the x-axis and on the y-axis (tricky stuff!)

The coordinates for the corners of the triangle are:

A: (–4,3)

B: (–4,–3)

C: (5,–3)

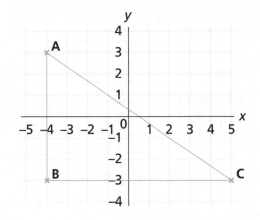

Have a go...

Explain to someone how you read the position of a point on a grid.

- *Draw a grid to help explain it.*
- *Include negative numbers on the axes.*

Key words

coordinates   quadrant

negative numbers

Quick Test

These are three corners of a square. What are the coordinates for:

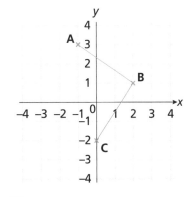

1. Position A
2. Position B
3. Position C
4. The fourth corner
5. If the square is translated so that A is moved to (0,2), where would position C now be?

Angles

Types of angles

An **angle** is a **measure of turn**. Angles are measured in **degrees (°)**.

There are 360° in a full circle.

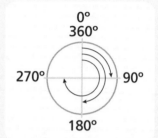

These are special angles to remember:

180° (straight line)   90° (right angle)   **acute** angle (less than a right angle)   **obtuse** angle (between 90° and 180°)   **reflex** angle (between 180° and 360°)

Measuring angles

A protractor is used to measure the size of an angle. It is a good idea to estimate the angle first and then measure it.

Be careful! Make sure you put the 0° line at the start position and read from the correct scale.

Read from 0 on the outer scale.

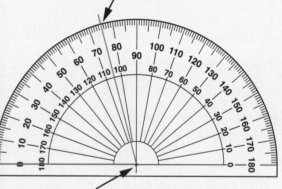

Count the degree lines carefully. This angle is 75.

Place the cross at the point of the angle you are measuring.

I like the idea of a cute angle – all soft and cuddly.

I think that you've forgotten that an acute angle is the sharpest kind – not very cuddly!

Angles and shapes

It is very useful to remember that:

All the angles in a triangle add up to 180°.

a + b + c = 180°

All the angles in a quadrilateral add up to 360°.

a + b + c + d = 360°

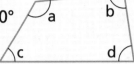

Angles and lines

Try to remember these angle facts:

Angles on a straight line add up to 180°.

Angles at a point add up to 360°.

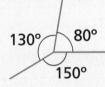

Perpendicular lines meet at 90°.

When two lines cross, the opposite angles are equal.

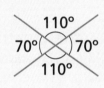

Have a go...

• Explain to someone how to measure angles with a protractor.

• Explain the different angle rules you can remember.

Key words

angle · obtuse
acute · reflex

Top Tip: Questions often have triangles with one angle missing. Find the total of the angles given and take it away from 180°.

Quick Test

1. Is this angle acute or obtuse?

2. What is the missing angle?

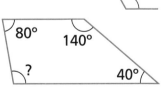

3. What is the size of angle A?

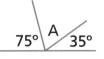

Test your skills

Robot spirals

Super Robot leaves a trail wherever he goes. He only travels along straight lines and always turns clockwise through 90°.

Here is his three-move trail for (1, 3, 2).

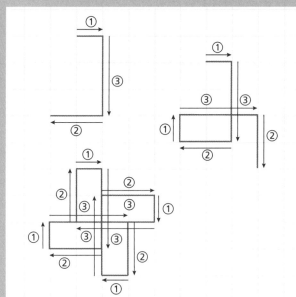

He repeats this move until he comes back to the starting point.

Draw these paths on squared paper.

(1, 4, 2)

(2, 5, 2)

(1, 1, 6)

Investigate other trails for three moves.

More robot spirals

If Super Robot makes four moves he goes a bit loopy – and runs off the page.

Here is his four-move trail for (1, 3, 2, 1).

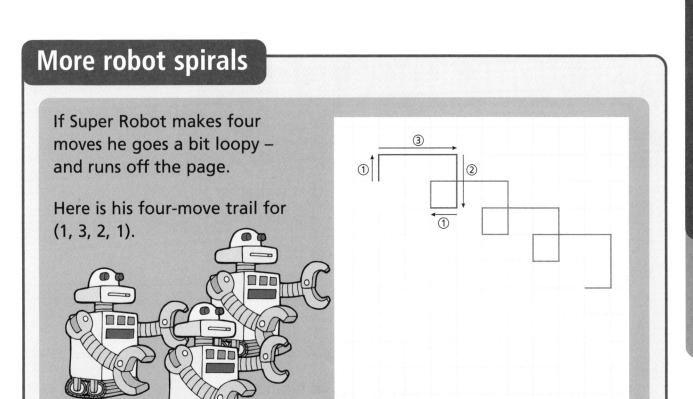

Investigate!

Investigate other trails for four moves.

What if....
- Super Robot made 5 moves, or 6, 7…
- He moved on a triangular grid

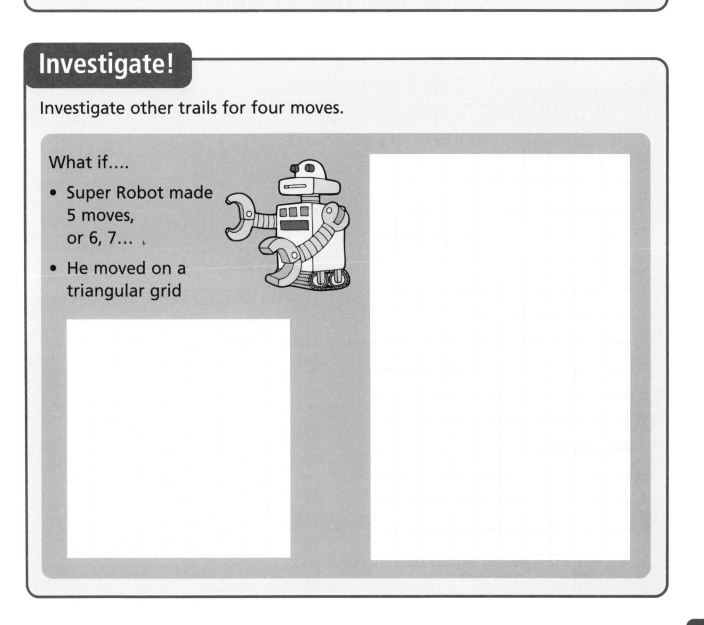

Test your knowledge

Section 1

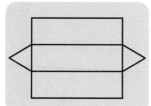

1 Name the shape that is made from this net. _____

2 How many vertices does this shape have? _____

3 Name this shape. _____

4 How many lines of symmetry does it have? _____

5 This shape is a regular polygon. True or false? _____

Section 2

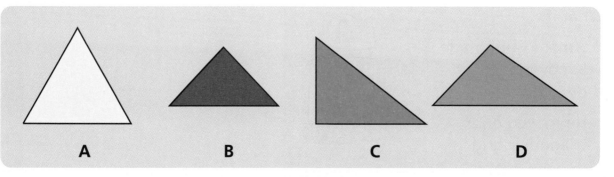

1 Write the correct letter to match each triangle to its name:

right-angled ____

isosceles ____

equilateral ____

scalene ____

2 Which triangle has 3 lines of symmetry? _____

3 Which triangle has an obtuse angle? _____

Section 3

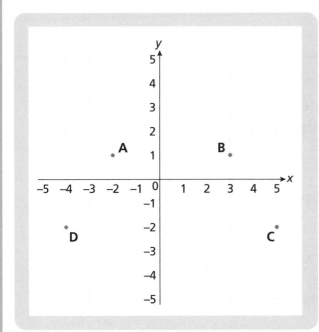

1. What are the coordinates for the positions A, B, C and D?

   _____

2. Name the shape when points A, B, C and D are joined in order.

   _____

3. What is the total of all the interior angles of this shape?

   _____

Section 4

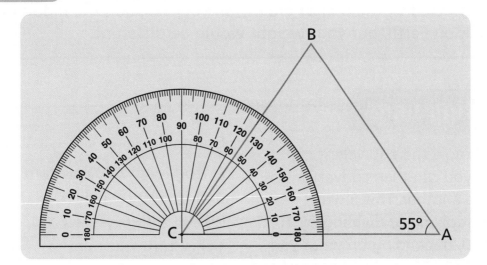

1. Read the size of angle C on the protractor. _____

2. Calculate the size of angle B. _____

3. Angle A is an acute angle. True or false? _____

4. What is the name of this type of triangle? _____

Measures

Equivalent measures

Length, mass (or weight) and capacity are all measured using different units. There is no getting away from the fact that you just need to know these **equivalent measures**:

Length

1 centimetre (cm) = 10 millimetres (mm)

1 metre (m) = 100 cm

1 kilometre (km) = 1000 m

Mass

1 kilogram (kg) = 1000 grams (g)

1 tonne = 1000 kg

Capacity

1 litre (l) = 1000 millilitres (ml)

1 centilitre (cl) = 10 ml

You may be asked questions about mass and weight. They are connected, but weight is affected by the pull of gravity, so your mass on the moon would be the same as on Earth, but your weight would be different.

Converting units

Once you know the equivalent measures, you can then convert from one unit to another. **This always means multiplying or dividing by 10, 100 or 1000**, depending on what you are converting. If you are not sure whether to multiply or divide – decide which one makes the most sense.

Remember, Milli means one thousandth, Centi means one hundredth, Kilo means one thousand.

Example

A frog leapt 120 cm from one lily pad to another. What is this distance in metres?

1 m = 100 cm

So, it is either 120 × 100 = 12 000 m or 120 ÷ 100 = 1.20 m

12 000 m is a mighty leap for a frog, so the obvious answer is:

120 cm = 1.20 m

Imperial measures

In the past, we used imperial measures. We still sometimes use **pints, gallons, pounds, inches and feet**, so it is worth knowing these:

Remember that ≈ means "is **approximately** equal to".

Length	Mass	Capacity
12 inches = 1 foot	16 ounces = 1 pound (lb)	8 pints = 1 gallon
2.5 cm ≈ 1 inch	25 g ≈ 1 ounce	1.75 pints ≈ 1 litre
30 cm ≈ 1 foot	2.25 lb ≈ 1 kg	4.5 litres ≈ 1 gallon
3 feet ≈ 1 metre		

Reading scales

A scale is the marking of lines to help us measure, e.g. up the side of a jug, on weighing scales or on a ruler. You need to read them carefully, using these steps:

1. Look at the unit – is it ml, cm, mm, g…?
2. If it is between numbers, work out what each mark means and count on or back.
3. If it is level with a number, read off that number.
4. Remember to include the unit of measurement (cm, mm, g) in the answer.

Have a go...

- *Take some items out of kitchen cupboards and estimate their mass and capacity.*
- *Check by reading the container and then check by measuring.*
- *Put the items back in the correct place!*

Key words

approximately

Quick Test

1. What is the mass in grams of a kitten weighing 2.3 kg?
2. A car petrol tank holds 10 gallons. About how many litres does it hold?
3. Which is greater, 650 cm or 7 metres?
4. What is the mass of this parcel?
5. How much water is in this jug?

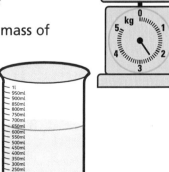
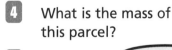

Time

Time facts

These are facts that you just need to know. Cover each one up and see what you can remember. Use a 'knuckle method' to learn the number of days in the months:

1 minute = 60 seconds
1 hour = 60 minutes
1 day = 24 hours
1 week = 7 days
1 fortnight = 14 days
1 year = 12 months = 365 days
leap year = 366 days

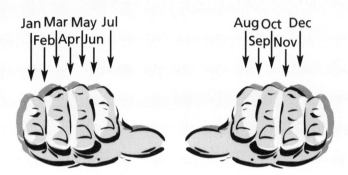

31 days : January, March, May, July, August, October, December.

All the 'knuckle months' have 31 days. February has 28 days (29 days in a leap year) and April, June, September and November have 30 days.

Telling the time

We read the time using either an **analogue clock** (with hands) or a **digital clock** (time in numbers). Digital clocks can be 12-hour or 24-hour clocks.

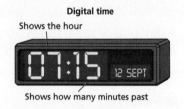

Both these clocks show seven fifteen or 15 minutes past 7.

Top Tip: am stands for 'ante meridiem' and means morning – from 12 midnight to 12 noon. pm stands for 'post meridiem' and means afternoon – from 12 noon to 12 midnight. Test questions often ask you to convert between 12-hour and 24-hour times. Look for the am and pm because that tells you if it is past midday.

24-hour time

Instead of using am and pm, **24-hour time goes from 00:00 to 24:00**.
So 3:50 am is 03:50 and 3:50 pm is 15:50.

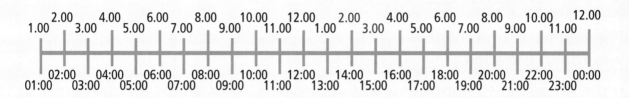

To read 24-hour time, am times look the same, but you add 12 hours to pm times. You always use 4 digits when you write the 24-hour clock, even for morning times. So 9.45 am is 09:45.

At midnight, a 24-hour clock shows 00:00. You probably did not know that because you are always asleep then (I hope!).

Calculating times

Test papers often have questions like this: A 'fun-filled family film' called 'Tongue-twister' starts at 6:40 pm and finishes at 8:15 pm. How long does the film last?

With any time questions like this, jot down a simple time line.

6:40 ⟶ 7:00 ⟶ 8:00 ⟶ 8:15
  20 mins   1 hr   15 mins

1 hr + 20 mins + 15 mins is **1 hr 35 mins**

Have a go...

Ask someone...

- *to give you a time. Convert it to 24-hour time*
- *to give you two times. Work out the difference between them.*

Key words

analogue clock   digital clock

Quick Test

1. How many days are there in August?
2. What is 7.40 pm using the 24-hour clock?
3. What is 22:25 using am or pm?
4. A family goes on a car journey that lasts for 2 hours 25 minutes. If they arrived at 5.10 pm, what time did they set off?
5. What time does this clock show?

Area and perimeter

Finding perimeters

The **perimeter** of a shape is easy to work out. It is just the distance all the way round the edge.

If the shape has straight sides, **add up the lengths of all the sides**. These may be given, or you may need to measure carefully along each of the sides using a ruler.

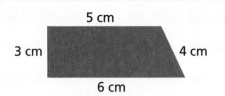

3 cm + 5 cm + 4 cm + 6 cm = 18 cm

The perimeter of this shape is 18 cm.

If it has curved sides, a piece of string or cotton may be useful. Go around the edge of the shape and then measure the length of the piece of thread.

Calculating perimeters

The perimeters of rectangles can be found using a simple formula:

2(a + b)

This means two times the sum of the length of the short and long sides.

Try it with these rectangles:

The perimeter of a square is simply 4 times the length of one of the sides.

Perimeter = 4 × 6 cm = 24 cm

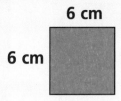

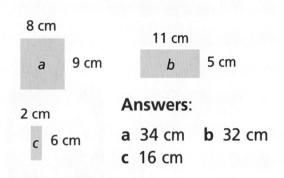

Answers:
a 34 cm b 32 cm
c 16 cm

Finding areas

The **area** of a shape is the amount of surface that it covers.

These shapes both have an area of 8 squares.

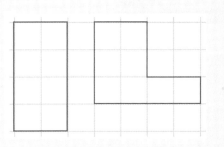

 Top Tip — If the shape has curved sides, count all the squares that are bigger than a half.

Areas of rectangles and shapes

Finding the area of a rectangle is easy if you know the length and width:

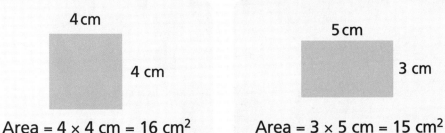

Area = 4 × 4 cm = 16 cm²

Area = 3 × 5 cm = 15 cm²

Area is usually measured in square centimetres, which is written as cm². Always remember to write this at the end of the measurement.

You might be asked to find the area of a shape that is made up from different rectangles joined together. **Just find the area of each part and then add them together**.

Area of big rectangle is 4 × 2 cm = 8 cm²

Area of square is 2 × 2 cm = 4 cm²

Total area = 12 cm²

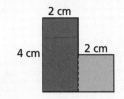

Have a go...

- Ask a friend to draw some shapes made from different rectangles.
- Work out the perimeter and area of each shape.

Key words

perimeter   area

Quick Test

What are the perimeter and area of these three shapes?

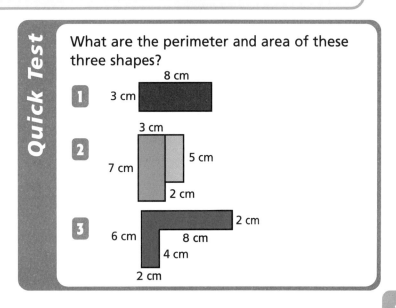

AREA AND PERIMETER MEASURES

Test your skills

Garden plan

This is a plan of Mr Broccoli's garden.

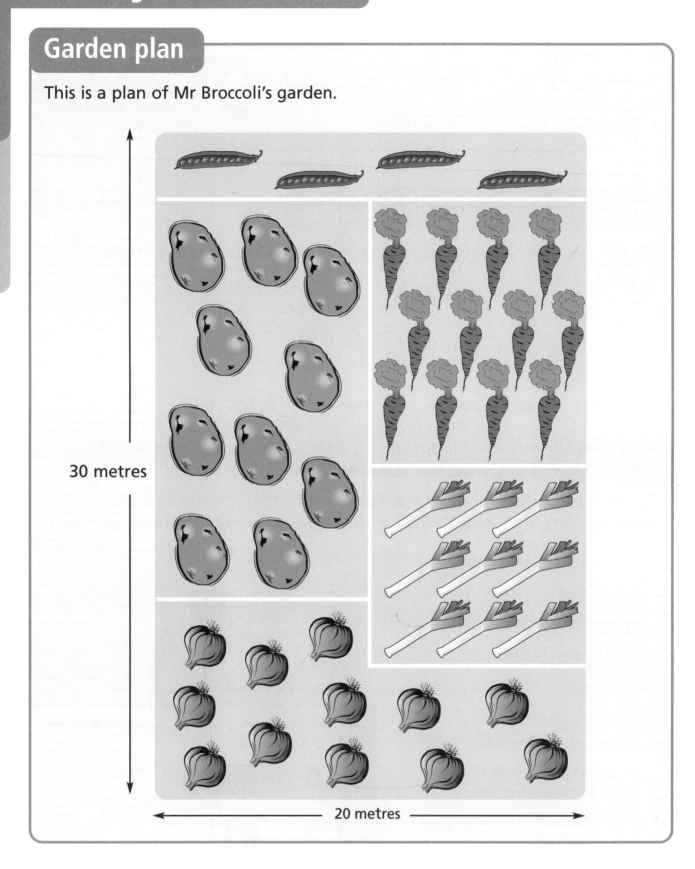

Garden percentages

This is how he divides his garden:

30% potatoes
25% onions
20% carrots
15% leeks
10% peas

Work out:

1. The area of the vegetable garden (m^2).
2. The area of the garden taken up by:

   potatoes   leeks
   onions   peas
   carrots

Garden information

Design your own vegetable garden with this information:

Area of vegetable garden: 450 m^2

Vegetables: 40% potatoes   12% beans   20% onions   10% carrots   18% peas

Garden design

Draw a plan of your garden.

Calculate the area taken up by each vegetable.

Test your knowledge

Section 1

1. A runner ran 1.6 km. How many metres did he run? _____

2. Approximately how many pints are there in 2 litres? _____

3. Which is heavier, 4500 g or 4 kg? _____

4. A jug holds 1.5 litres. If 800 ml of water is poured out from a full jug, how much water is left in the jug? _____

5. An equilateral triangle has sides of length 25 mm. What is the length of the perimeter of the shape, in centimetres? _____

Section 2

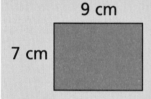

1. Calculate the area of this rectangle. _____

2. What is its perimeter? _____

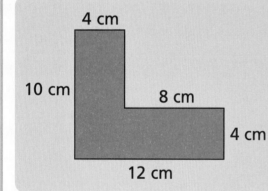

3. Calculate the area of this shape. _____

4. What is its perimeter? _____

Section 3

1 How many days are there in June? _____

2 What is 14:36 using am or pm? _____

3 A film is 1 hour 15 minutes long. If it starts at 7.20 pm, when will it finish? _____

	Bus 1	Bus 2
Bus station	0950	1220
Shopping centre	1012	1239
Hospital	1025	1255
Park	1034	1304

4 How long does it take Bus 1 to travel from the bus station to the hospital? _____

5 Which bus is quicker from the shopping centre to the hospital? _____

6 Which bus takes 25 minutes to travel from the shopping centre to the park? _____

Section 4

1 A square has a perimeter of 20 cm. What is its area? _____

2 How many 200 ml jugs can be filled from a 5 litre jug? _____

3 What is the difference in weight between two parcels weighing 850 g and 1.2 kg? _____

4 How many millimetres are there in 25 centimetres? _____

5 How many days are there in April? _____

6 The perimeter of a square garden is 248 metres. What is the length of each side? _____

Probability

Chance and probability

Without realising, you have probably already used some **probability language** today. It is used for everyday events and is also important in maths.

Examples

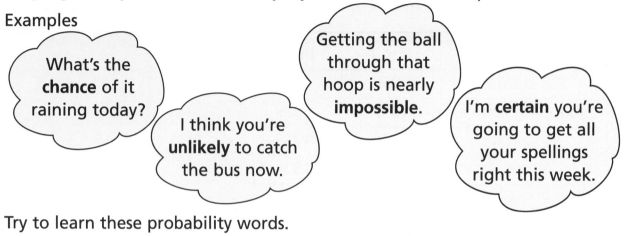

Try to learn these probability words.

The probability scale

In maths, probability is a little more accurate than the examples above. **A probability scale can be used to show how likely an event is to happen:**

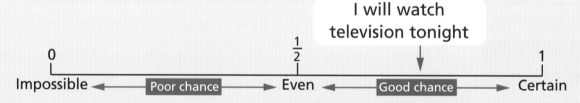

Where do you think these statements will be on the scale?

- I will break a world record this year.
- I will grow taller than my mum.
- Tomorrow will be Thursday.
- Cress will grow with no water.

Even chance

Even chance is an **equal chance of something happening as not happening**. We also say a **1 in 2 chance** or a **50:50 chance**.

Examples Tossing a coin – there is an even chance of it landing on heads.
Rolling a dice – there is an even chance that it will land on an odd number.

Bead experiments

Look at this bag of beads.

What is the probability of picking out a red bead?

There are 12 beads and 6 of them are red.

$\frac{6}{12}$ is the same as $\frac{1}{2}$ so there is a 1 in 2, or an even chance of picking out a red bead.

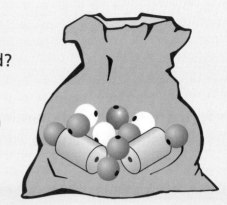

What is the probability of picking out:
- a blue bead?
- a yellow bead?

There is a 1 in 6 or $\frac{1}{6}$ chance of picking out a blue bead and a 1 in 4 or $\frac{1}{4}$ chance of picking out a yellow bead.

Using a dice

Dice experiments are useful for testing probabilities.

On a 1–6 dice, what is the probability of throwing:
- a six?
- an even number?
- a number smaller than 3?
- a number greater than 7?
- a number between 0 and 7?

These are the answers muddled up. Try to match them.

$1, \frac{1}{3}, \frac{1}{2}, \frac{1}{6}, 0$

Remember, 0 is impossible – there needs to be absolutely no chance of it happening, and 1 is certain – it will absolutely, definitely happen. Most events lie in between these two extremes.

Have a go...

Use the number cards from a pack of playing cards. Work out the probabilities of picking out:
- *a red card (hearts and diamonds)*
- *an odd number*
- *a multiple of 5*
- *a diamond.*

Carry out an experiment by shuffling and repeatedly picking out cards, to see how true the probabilities are.

Key words

probability

Quick Test

Look at this spinner. What is the chance of spinning:

1. blue?
2. red?
3. either red or blue?
4. green?
5. yellow?
6. red, blue or green?

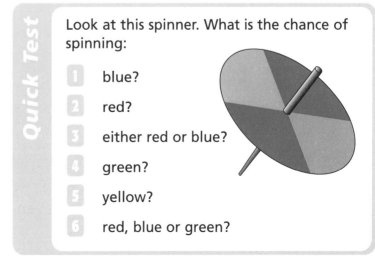

PROBABILITY DATA

Charts and graphs

Bar charts

Information can be shown in many ways using different charts and graphs. To understand bar charts, remember:

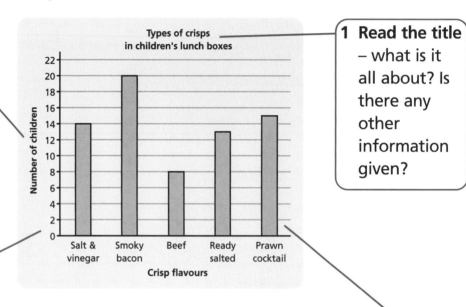

1 Read the title – what is it all about? Is there any other information given?

2 Look at the axis labels – these should explain the horizontal and vertical lines.

3 Work out the scale – look carefully at the numbers – do they go up in 1s, 2s, 5s, 10s…?

4 Compare the bars – read them across to work out the amounts.

Pie charts

These are **circles divided into sections**. Each section represents a number of items.

Read the questions carefully. **You could be asked to give a fraction, a percentage or a number as an answer.**

A survey of 24 children was carried out to find their favourite types of sandwich. This pie chart shows the results:

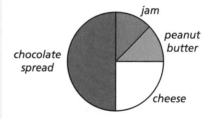

$\frac{1}{4}$ of the children chose cheese as their favourite sandwich and there were 24 children altogether.

So $\frac{1}{4}$ of 24 is 6 children choosing cheese.

Top Tip *Always look at the total for the whole pie, and then work out what each section is worth by seeing what fraction of the pie it represents.*

Frequency charts

The word frequency just means 'how many', so **a frequency chart is a record of how many there are in a group**. This is nice and simple when single items can be compared. Look at the barline chart.

A barline graph is just like a bar chart, but the bars are thinner!

Frequency charts with **grouped data** are a bit trickier, but very useful.

Imagine you wanted to compare the lengths of the tails of 15 dogs in a dogs' home (silly idea I know!) to find the most common length.

Most of the tails would be different lengths, so you may end up with a graph a bit like the one on the right.

It would be much more useful to group the lengths and then compare them:

So the most common length is between 21 cm and 30 cm.

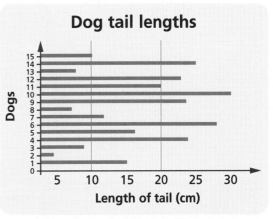

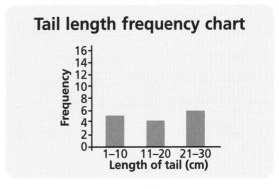

Have a go...

Carry out a survey of shoes worn by your family. Draw graphs to show your results. They could show:

- *types of shoes (trainer, work, school, walking...)*
- *fastenings (lace, Velcro, slip-on...)*
- *weight and size.*

Key words

axis

Quick Test

Use the graphs and charts to answer these:

1. Which flavour of crisp was found most often in children's lunch boxes?
2. How many times did United score 4 goals in a match?
3. How many children chose peanut butter as their favourite sandwich?
4. How many dogs had tails between 21 and 30 cm in length?

Line graphs

Reading line graphs

A graph with numbers for both **axes** is often shown as a line. Reading line graphs is easy if you follow these three simple steps:

1. **Read all the information given** so you understand what the graph is about.
2. Use a ruler to **go up from the horizontal axis to meet the line.**
3. From this point **read across to the vertical axis to give the value.**

Conversion graphs

Straight-line graphs are often used to **show one amount converted to another**.

For example, if you wanted to change litres into pints, this graph will make it easy.

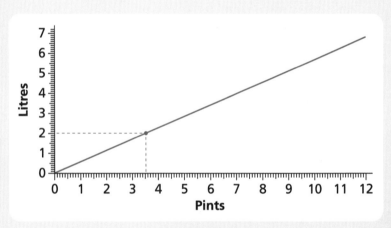

Converting currency is also easy with a straight-line graph.

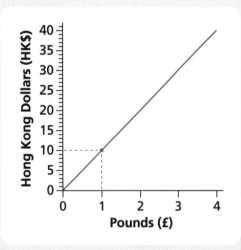

This graph shows the conversion between Hong Kong Dollars (HK$) and British Pounds (£).

If you need to draw a conversion graph it is important to get the scales right. So for this one, the Hong Kong Dollars can go up in fives and the pounds in ones. Once you know that there are 10HK$ for each £1, you can plot the 10HK$ against £1 and join this with a straight line to zero. Different values can then be read off.

Time/distance graphs

Time/distance graphs are exactly what they say – **they look at the time taken for each part of a journey**.

This is a record of the Mitchell family's car journey lasting 4 hours.

This horizontal line shows that no distance was travelled for 30 minutes.

The steeper the line, the faster the journey.

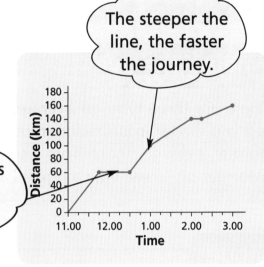

Remember to use the three steps for reading line graphs.

Joining crosses

Many line graphs show a set of points or crosses joined by a line. This often means that **the lines between the crosses do not show any real values, but a trend or an idea of how things are changing**.

This line graph shows the monthly sales of 'Bendy Soft Toy' products (rounded to the nearest 100 sales).

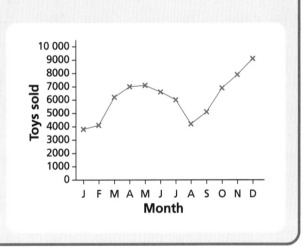

Have a go...

When you go on a car journey, or any long journey, keep a chart of the times and distance travelled. When you get home, draw it up as a time/distance graph.

Quick Test

Use the graphs and charts to answer these:

1. How many litres are approximately the same as 10 pints?
2. How many pints are approximately equivalent to 4 litres?
3. How many Hong Kong Dollars are worth £1.50?
4. At what time had the Mitchell family travelled 140 km?
5. How far had they travelled by 1.00 pm?
6. How many Bendy toys were sold in May?

Key words

axes

Averages

Finding the average

There are three types of average – mode, median and mean. They all try to tell you similar information in different ways – the middle, most common, normal, average figure from a set. Once you know this then you can compare information to see which figures are above or below average.

Mode

Look at this set of shoe sizes for Snow White and the Seven Dwarfs. Snow White is the one with unusually large feet!

The mode of a set of data is **the one that occurs the most often**. The modal average for these shoe sizes is size 2 – because there are more of this size than any other.

Top Tip: Rearrange the numbers and list them in order of size to help work out any type of average.

Mean

This is what we normally think of as average: the mean average.

mean = total ÷ the number of items

You can remember mean because it is the meanest average to try to work out!

This table shows the number of golden eggs laid by the hen owned by Jack (of beanstalk fame). What is the mean average number of eggs laid?

Day	Number of eggs laid
1	3
2	5
3	5
4	9
5	8

Mean = total ÷ number of items

$$\frac{3 + 5 + 5 + 8 + 9}{5} = \frac{30}{5} = 6$$

So the mean average is 6 eggs.

Median

The median is easy to remember – **it is the middle number**.

This chart shows the number of buckets of water Jack and Jill collected each day for a week (luckily not in the week that Jack fell down and broke his crown).

Monday	9 buckets
Tuesday	8 buckets
Wednesday	5 buckets
Thursday	6 buckets
Friday	9 buckets
Saturday	3 buckets
Sunday	4 buckets

To work out the median, follow these three easy steps:

1. **Put the numbers in order from smallest to largest**
   3, 4, 5, 6, 8, 9, 9

2. **Count how many numbers you have altogether**
   3, 4, 5, 6, 8, 9, 9 – there are seven numbers

3. **Go to the middle number**
   3, 4, 5, 6, 8, 9, 9 – the fourth number.

So the median for this set of numbers is 6 buckets of water.

The **range** tells us how much the information is spread between the smallest and the largest amount.

To find the range, take the smallest and largest digits. Range is from 3 to 9, which is 6 buckets.

When working out the median and there is an even amount of numbers, you take the two middle numbers, add them together and divide by two.

Did you know mode also means fashionable? A bit like me really.

And I thought you were just mean!

Have a go...

- Look in a newspaper and find the weather forecast.
- List the temperatures of 10 places.
- Find the mode, median and mean average temperatures.

Quick Test

These are the prices for nine different comics:
80p 70p 45p 90p 65p 90p 70p 50p 70p

1. What is the mode?
2. What is the median?
3. What is the mean?
4. What is unusual about this?

Key words

average	mean
mode	range
median	

Test your skills

Rolling dice

You are more likely to roll a 3 on a dice than any other number.

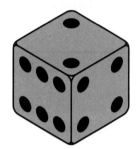

Is this true?

Roll a dice 100 times and record your results using a tally chart.

Use the tally chart to make a frequency chart using a computer database or on squared paper.

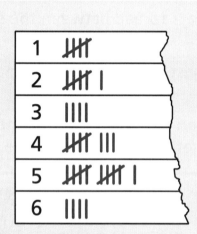
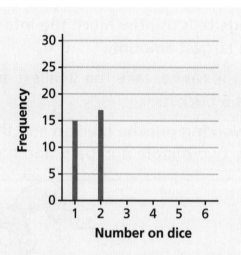

Analyse your results:

Did you expect the results to be like this? _____

Which number was rolled most often? _____

Do you think that the next time you roll a dice you are more likely to roll a 3 or a 6? Why?

_____
_____
_____

Rolling a pair of dice

Roll a pair of dice 100 times, totalling the pair each time you roll.

Which total do you think will be made the most?

_____

Record your results using a tally chart.

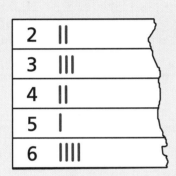

2			
3				
4			
5		
6					

7	ЖГ
8	ЖГ	
9					
10			
11				
12			

Use the results to make a frequency chart on a computer or on squared paper.

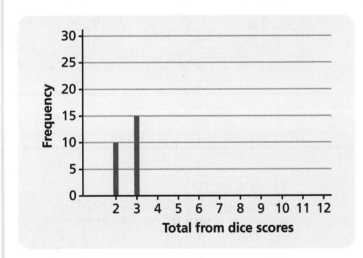

Which total is the least likely? _____

Which total do you think is the most likely to be made next time you roll? _____

Test your knowledge

Section 1

1. How many children travel to school by bus? _____
2. How many more children walk than cycle? _____
3. How many children are there in Y6 in total? _____

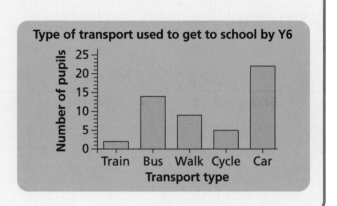

Section 2

1. How many cars were counted between 2 and 3 pm? _____
2. During which hour were 17 cars counted? _____
3. In which hour were most cars counted? _____

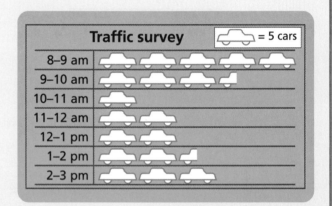

Section 3

1. What distance had the lorry completed by 9.30? _____
2. At what time had the lorry driven 80 km? _____
3. How long did the driver stop for lunch? _____
4. How far did the lorry travel in total? _____

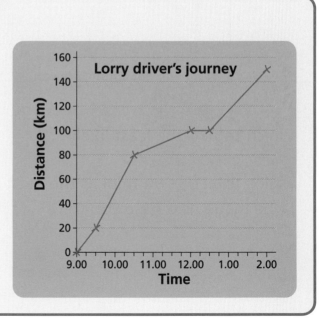

Section 4

1 How many people owned cats? _____

2 What fraction of the people owned a hamster? _____

3 How many people owned a guinea pig or rabbit? _____

4 A hamster weighs 2 oz. Approximately how much is this in grams? _____

5 Ben's rabbit weighs 14 oz. Approximately how many grams does it weigh? _____

6 Sanjay's guinea pig weighs 500 g. Approximately how many ounces is this? _____

A survey asked 48 pet owners about the types of pets they owned.

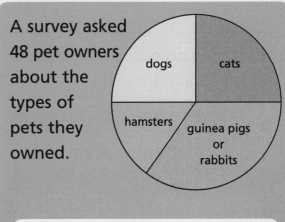

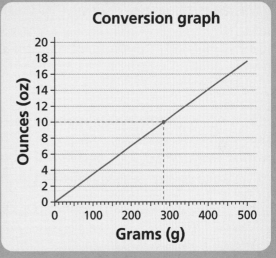

Section 5

These are the weights of seven guinea pigs.

| 470 g | 330 g | 550 g | 420 g | 440 g | 420 g | 450 g |

1 What is the mode? _____

2 What is the median? _____

3 What is the mean? _____

4 What is the range? _____

National Test practice

1 mark for each question

1. Write the number fourteen thousand, nine hundred and sixty-five. _____

2. Put these fractions in order, starting with the smallest:

   $2\frac{3}{4}, 4\frac{3}{4}, 4\frac{1}{4}, 3\frac{1}{4}$ _____ _____ _____ _____

3. Round 3118 to the nearest 10. _____

4. What is the next number in this sequence?

   11   6   1   –4   –9   _____

5. What is the difference between –4 and 2? _____

6. 5 is a factor of 3940. True or false? _____

7. In a bakery 1 kg of butter is mixed with every 3 kg of flour to make 4 kg of pastry. The bakery needs 16 kg of pastry every day. How much flour is needed for one day? _____

8. What is 4.2 × 100? _____

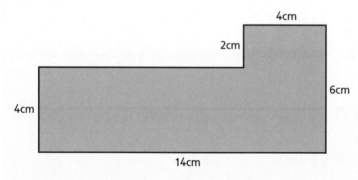

9. What is the area of this shape? _____

10. What is the perimeter? _____

11. A CD player costs £50. There is 20% off in the sale. What is the sale price? _____

12. 44 + 45 = ? _____

13. Place these decimals in order, starting with the smallest:

    3.7 3.05 3.78 3.3 3.5 3.23

    ____ ____ ____ ____ ____ ____

14 4.05 + 19.5 = ? _____

15 Dean has 34p. How many 4p sweets can he buy? _____

16 In a sponsored swim, Josh completed 441 m, 159 m further than Kate. How far did Kate swim? _____

17 What is the probability of throwing a six on a dice? _____

18 6381 − 457 = ? _____

19 What is 5.15 pm using the 24-hour clock? _____

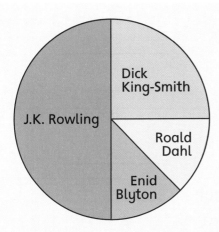

20 40 children were asked to name their favourite author. Which author was chosen by 20 children? _____

21 How many children said Roald Dahl was their favourite author? _____

22 What is the name of a shape with 3 equal sides and 3 equal angles? _____

23 How many faces are there on a cuboid? _____

24 What is $\frac{3}{5}$ as a decimal? _____

25 ? ÷ 8 = 23 _____

26 Place these numbers in the correct places:

473   374   437   _____ < _____ < _____

27 Which of these fractions is the same as one whole?

$\frac{3}{5}$ $\frac{3}{4}$ $\frac{4}{4}$ $\frac{4}{5}$ _____

28 What is the nearest number to 60 000 you can make from the digits 9 3 4 5 7? _____

3 6 4 4 2 3 5 4 4 6 4 4 3 3 5

These are the shoe sizes of fifteen 11 year-olds.

29 What is the mode? _____

30 What is the median? _____

31 Which is the larger number, 6.24 or 6.4? _____

32 There are 26 eggs and n eggs hatch. What is the equation that shows how many eggs are left? _____

33 What is $\frac{2}{3}$ of 21? _____

34 What is 15% as a fraction? _____

35 3 is a factor of 423. True or false? _____

36 What is the missing number in this sequence?

1   3   ?   10   15   21 _____

37 The temperature at 9 am was −2°C. By lunchtime it had risen by 8°C. What was the temperature at lunchtime? _____

38 3227 × 100 = ? _____

39 Name this shape. _____

40 This shape has rotational symmetry of order 5. True or false? _____

0	10	20	30	40
A bit more practice needed		Great try – check your errors		Fantastic – ready for the test!

Answers

Numbers

PAGES 4–5 PLACE VALUE
1. twenty-three thousand, six hundred and eight
2. 13 478
3. 189 600
4. 345
5. 10 084

PAGES 6–7 COMPARING AND ORDERING
1. 61 049
2. 348 < 384 < 438
3. 379, 485, 2044, 2114, 2611, 4205, 10 901
4. a F b T c F

PAGES 8–9 ROUNDING AND APPROXIMATION
1. 3460
2. 48 500
3. 686 000
4. 30 000
5. c

PAGES 10–11 NUMBER PATTERNS
1. a 14 b –8 c 195 d 80
2. 11
3. –3°C

PAGES 12–13 FORMULAE AND EQUATIONS
1. 10, 1, 16, 10
2. 20 – n
3. 2
4. c

PAGES 14–15 SPECIAL NUMBERS
1. 100
2. 15
3. 21, 19
4. 6
5. 121

PAGES 16–17 FACTORS AND MULTIPLES
1. 1, 20, 2, 10, 4, 5
2. yes
3. 92 and 96
4. true
5. 42
6. 19

PAGES 18–19 FRACTIONS
1. $\frac{7}{8}$
2. $\frac{2}{6}$
3. $\frac{2}{3}$
4. a $\frac{12}{20}$ b $\frac{3}{4}$ c $\frac{25}{30}$

PAGES 20–21 COMPARING FRACTIONS
1. $\frac{5}{8}$
2. $\frac{1}{10}, \frac{3}{10}, \frac{7}{10}, \frac{9}{10}$
3. $\frac{1}{3}, \frac{1}{2}, \frac{5}{9}, \frac{5}{6}$
4. 0.8
5. $\frac{5}{20}$

PAGES 22–23 DECIMALS AND FRACTIONS
1. 0.7
2. $\frac{2}{5}$
3. 28.60
4. 0.03
5. 0.2785

PAGES 24–25 COMPARING DECIMALS
1. 4.7
2. 6.08, 6.12, 6.3, 6.8, 6.95
3. 13
4. 18.6
5. 6.38

PAGES 26–27 RATIO AND PROPORTION
1. 12
2. $\frac{3}{5}$
3. 10
4. 960 g

PAGES 28–29 PERCENTAGES
1. 0.15
2. $\frac{1}{20}$
3. 80%
4. 60%
5. 1%

PAGES 32–33 TEST YOUR KNOWLEDGE

Section 1
1. twenty-seven thousand, three hundred and eighty-four
2. 86 210
3. 648
4. 51 001, 6117, 4308, 1751, 756, 300
5. 5000

Section 2
1. 489 > 420 > 240
2. 25
3. 7
4. 1, 2, 3, 5, 6, 10, 15 and 30
5. no
6. 7
7. true
8. 3730
9. top line: 4, 7 bottom line: 32, 8

Section 3
1. $\frac{2}{6}$
2. $\frac{3}{4}$
3. 0.4
4. $\frac{3}{10}$
5. 351
6. 1.72
7. 2.3 m, 3.44 m, 4.05 m, 423 m

Section 4
1. 5
2. $\frac{1}{3}$
3. 75%
4.

Decimals	0.5	0.1	0.3	0.4	0.25	0.6
Fractions	$\frac{1}{2}$	$\frac{1}{10}$	$\frac{3}{10}$	$\frac{4}{10}$	$\frac{1}{4}$	$\frac{3}{5}$
Percentages	**50%**	**10%**	30%	40%	**25%**	60%

Calculations

PAGES 34–35 MENTAL MATHS
1. 91
2. 55
3. 151
4. 25
5. 45
6. 38

PAGES 36–37 MULTIPLICATION FACTS
1. a 36 b 48 c 28 d 56
2. 204
3. a 23 b 78 c 13 d 12

PAGES 38–39 ADDITION AND SUBTRACTION
1. 612
2. 1135
3. 3582
4. 356
5. 177

PAGES 40–41 MULTIPLICATION METHODS
1. 612
2. 4800
3. 5208
4. 8256
5. 21 164

PAGES 42–43 DIVISION METHODS
1. 17
2. 53.7
3. £77
4. 35
5. 46

PAGES 44–45 FRACTIONS OF QUANTITIES
1. 2
2. 20
3. £1.25
4. 30 kg
5. 8

PAGES 46–47 PERCENTAGES AND MONEY
1. £6
2. £30
3. £7.50
4. £9.24

PAGES 48–49 PROBLEMS
1. 360
2. 28
3. 5
4. £1.40

PAGES 52–53 TEST YOUR KNOWLEDGE

Section 1
1. 82
2. 122
3. 39
4. 53
5. 378
6. 27
7. 171
8. £24

Section 2
1. a 5289 b 338 c 1828
2. a 1829 b 78 c 60.17

Section 3
1. £37.50
2. £91
3. 12 years old
4. 29 ml

Section 4
1. 22
2. 312
3. 1970 g
4. 44p
5. £1
6. yes

Section 5
1. 75%
2. 45
3. £56
4. 72 g
5. 15

Shapes

PAGES 54–55 2-D SHAPES
1. pentagon
2. they have equal sides and angles
3. isosceles
4. 24 cm
5. equilateral

PAGES 56–57 3-D SOLIDS
1. square-based pyramid
2. corner
3. 6
4. cuboid
5. 8

PAGES 58–59 GEOMETRY
1. 1
2. 2
3. true
4. reflected

PAGES 60–61 USING COORDINATES
1. (−1,3)
2. (2,1)
3. (0,−2)
4. (−3,0)
5. (1,−3)

PAGES 62–63 ANGLES
1. obtuse
2. 100°
3. 70°

PAGES 66–67 TEST YOUR KNOWLEDGE

Section 1
1. triangular prism
2. 6
3. hexagon
4. 6
5. true

Section 2
1. right-angled C, equilateral A, isosceles B, scalene D
2. A
3. D

Section 3
1. A (−2,1) B (3,1) C (5,−2) D (−4,−2)
2. trapezium
3. 360°

Section 4
1. 55°
2. 70°
3. true
4. isosceles

Measures

PAGES 68–69 MEASURES
1. 2300 g
2. 45
3. 7 metres
4. 2.4 kg
5. 650 ml

PAGES 70–71 TIME
1. 31
2. 19:40
3. 10.25 pm
4. 2.45 pm
5. 8.40

PAGES 72–73 AREA AND PERIMETER
1. area = 24 cm², perimeter = 22 cm
2. area = 31 cm², perimeter = 24 cm
3. area = 28 cm², perimeter = 32 cm

PAGES 76–77 TEST YOUR KNOWLEDGE

Section 1
1. 1600
2. 3.5
3. 4500 g
4. 700 ml
5. 7.5 cm

Section 2
1. 63 cm²
2. 32 cm
3. 72 cm²
4. 44 cm

Section 3
1. 30
2. 2.36 pm
3. 8.35 pm
4. 35 minutes
5. Bus 1
6. Bus 2

Section 4
1. 25 cm²
2. 25
3. 350 g
4. 250
5. 30
6. 62 m

Data

PAGES 78–79 PROBABILITY
1. 1 in 4 ($\frac{1}{4}$)
2. 1 in 2 ($\frac{1}{2}$)
3. 3 in 4 ($\frac{3}{4}$)
4. 1 in 4 ($\frac{1}{4}$)
5. 0
6. 1

PAGES 80–81 CHARTS AND GRAPHS
1. smoky bacon
2. 9
3. 3
4. 6

PAGES 82–83 LINE GRAPHS
1. 5.7
2. 7
3. 15
4. 2.00 pm
5. 100 km
6. 7000

PAGES 84–85 AVERAGES
1. 70p
2. 70p
3. 70p
4. they are all the same

PAGES 88–89 TEST YOUR KNOWLEDGE

Section 1
1. 14
2. 4
3. 52

Section 2
1. 15
2. 9–10 am
3. 8–9 am

Section 3
1. 20 km
2. 10.30 am
3. 30 minutes
4. 150 km

Section 4
1. 12
2. $\frac{1}{8}$
3. 18
4. 56 g
5. 400
6. 17.5

Section 5
1. 420 g
2. 440 g
3. 440 g
4. 330 g – 550 g

National Test practice

1. 14 965
2. $2\frac{3}{4}$, $3\frac{1}{4}$, $4\frac{1}{4}$, $4\frac{3}{4}$
3. 3120
4. −14
5. 6
6. true
7. 12 kg
8. 420
9. 64 cm²
10. 40 cm
11. £40
12. 89
13. 3.05, 3.23, 3.3, 3.5, 3.7, 3.78
14. 23.55
15. 8
16. 282 m
17. 1 in 6 or $\frac{1}{6}$
18. 59 24
19. 17:15
20. J.K. Rowling
21. 5
22. equilateral triangle
23. 6
24. 0.6
25. 184
26. 374 < 437 < 473
27. $\frac{4}{4}$
28. 59 743
29. size 4
30. size 4
31. 6.4
32. 26 − n
33. 14
34. $\frac{3}{20}$
35. true
36. 6
37. 6°C
38. 322 700
39. pentagon
40. true

Glossary

acute an angle smaller than a right-angle; between 0° and 90°

adjacent near or next to something

analogue clock shows the time using hands moving around a dial

angle the amount by which something turns; measured in degrees (°)

anti-clockwise turning in this direction, opposite to the hands of a clock

approximate an approximate answer is not exactly but very close to the right answer; the sign ≈ means 'is approximately equal to'

area the area of a shape is the amount of surface that it covers

average the middle or most common amount. There are three types of average: mode, median and mean

axis (plural is axes) the horizontal and vertical lines on a graph

circumference the distance all the way around the outside of a circle

clockwise turning in this direction, like the hands of a clock

coordinates numbers that give the position of a point on a graph or grid

decimals any number made up of the digits 0 to 9

denominator the bottom number of a fraction, the number it is divided into. Example: $\frac{2}{3}$

diameter the distance right across the middle of a circle

difference the amount by which one number or value is greater than another

digit there are 10 digits: 0, 1, 2, 3, 4, 5, 6, 7, 8 and 9 that make all the numbers we use

digital clock shows the time using digits rather than by having hands on a dial

edge where two faces of a 3-D shape meet

equation where symbols or letters are used instead of numbers. Example: 3y = 12, so y = 4

equivalent fractions these are equivalent fractions: $\frac{1}{2} = \frac{2}{4} = \frac{3}{6}$

estimate like guessing, only using information to get an, approximate answer

even number a number that can be divided exactly by 2. Even numbers end in 0, 2, 4, 6 or 8

face the flat side of a 3-D shape

factor a number that will divide exactly into other numbers. Example: 5 is a factor of 20

formula a formula (plural is formulae) uses letters or words to give a rule

HTU hundreds, tens and units

inverse opposite in effect. For example, the inverse of x is –x

mean this is the total divided by the number of items. Example: the mean of 3, 1, 6 and 2 is (3 + 1 + 6 + 2) ÷ 4 = 3

median the middle number in an ordered list. Example: 3, 8, 11, 15, 16. The median number is 11

million the number 1000 000; it is one thousand thousand

mode the most common number in a list. Example: 2, 6, 4, 2, 5, 5, 2. The mode is 2

multiple a multiple is a number made by multiplying together two other numbers

negative number a number less than zero on the number line

net the net of a 3-D shape is what it looks like when it is opened out flat

numerator the top number of a fraction. Example: $\frac{3}{5}$

obtuse an angle greater than a right-angle but smaller than a straight line, so between 90° and 180°

odd numbers numbers that cannot be divided exactly by 2. Odd numbers always end in 1, 3, 5, 7 or 9

parallel lines that are parallel never meet

percentage this is a fraction out of 100, shown with a % sign. Example: 50% shows $\frac{50}{100}$ or $\frac{1}{2}$

perimeter the distance all the way around the edge of a shape or object

polygon any straight-sided flat shape

positive number a number greater than zero on the number line

prime number any whole number, apart from 1, that can only be divided by itself and by 1, without leaving a remainder

probability the chance or likelihood that something will happen

proportion this is the same as finding the fraction of the whole amount. Example: the proportion of red cubes is 3 out of 5 or $\frac{3}{5}$

quadrant one-quarter of a circle

radius distance from the centre of a circle to the edge

range the spread of data. It is the difference between the greatest and least values

ratio compares one amount with another. Example: the ratio of red cubes to blue cubes is 3:2

reflex an angle measuring between 180° and 360°

rounding rounding a whole number means to change it to the nearest ten, hundred or thousand to give an approximate number; decimal numbers can be rounded to the nearest whole number, tenth or hundredth

sequence a list of numbers that usually has a pattern

square number numbers multiplied by themselves make square numbers. Example: 4 × 4 = 16. The first five square numbers are 1, 4, 9, 16 and 25.

square root the opposite of a square number; a number, when multiplied by itself, makes a square number. Example: the square root of 25 is 5

symbol a letter or sign that represents a specific quantity or function

symmetrical a symmetrical shape is one that is balanced about a point, line or plane

thousand the number 1000; it is 10×100

3-dimensional a solid shape is three-dimensional because it has length, width and height

triangular numbers numbers made by triangle patterns. Example: 1 + 2 = 3, 1 + 2 + 3 = 6. The first five triangular numbers are 1, 3, 6, 10 and 15

Venn diagram a diagram that shows groups of things by putting circles around them

vertex (plural is vertices) this is the corner of a 3-D shape, where edges meet